SpringerBriefs in Electrical and Computer Engineering

Series Editors

Tamer Başar, Coordinated Science Laboratory, University of Illinois at Urbana-Champaign, Urbana, IL, USA

Miroslav Krstic, La Jolla, CA, USA

SpringerBriefs present concise summaries of cutting-edge research and practical applications across a wide spectrum of fields. Featuring compact volumes of 50 to 125 pages, the series covers a range of content from professional to academic. Typical topics might include: timely report of state-of-the art analytical techniques, a bridge between new research results, as published in journal articles, and a contextual literature review, a snapshot of a hot or emerging topic, an in-depth case study or clinical example and a presentation of core concepts that students must understand in order to make independent contributions.

More information about this series at http://www.springer.com/series/10059

Mitul Kumar Ahirwal • Anil Kumar
Girish Kumar Singh

Computational Intelligence and Biomedical Signal Processing

An Interdisciplinary, Easy and Practical Approach

Mitul Kumar Ahirwal
Department of Computer Science
and Engineering
Maulana Azad National Institute
of Technology
Bhopal, Madhya Pradesh, India

Anil Kumar
Department of Electronics and Communications
Engineering
PDPM Indian Institute of Information
Technology Design and Manufacturing
Jabalpur, Madhya Pradesh, India

Girish Kumar Singh
Department of Electrical Engineering
Indian Institute of Technology Roorkee
Roorkee, Uttarakhand, India

ISSN 2191-8112   ISSN 2191-8120   (electronic)
SpringerBriefs in Electrical and Computer Engineering
ISBN 978-3-030-67097-9   ISBN 978-3-030-67098-6   (eBook)
https://doi.org/10.1007/978-3-030-67098-6

This Springer imprint is published by the registered company Springer Nature Switzerland AG
The registered company address is: Gewerbestrasse 11, 6330 Cham, Switzerland

This book is dedicated to our parents and family members.

Foreword

This book is written by three authors having expertise in different domains. Dr. Mitul Kumar Ahirwal is working with the Department of Computer Science and Engineering, Maulana Azad National Institute of Technology, Bhopal, MP, India, and has expertise in computational intelligence techniques and MatLab. Dr. Anil Kumar is working with the Department of Electronics and Communications Engineering, PDPM Indian Institute of Information Technology Design and Manufacturing, Jabalpur, MP, India, and has expertise in digital signal processing and digital filter design for biomedical signals. Dr. Girish Kumar Singh is working in the Department of Electrical Engineering, Indian Institute of Technology Roorkee, UK, India, and has expertise in digital signal processing, biomedical signals, and healthcare systems. He also has more than 25 years of teaching experience in India and abroad.

This book is a very good choice for readers who wish to continue their studies and research in interdisciplinary areas of computation intelligence and biomedical signal processing. The authors of this book are also from different domains, and they have provided their practical experiences in the form of examples and studies. They have also explained all the topics in a very simplified manner to provide a clear understanding of each topic and its uses and applications. That is why this book will be extremely useful for beginners having an interest in computation intelligence and biomedical signal processing field.

Department of Electronics and
Communication Engineering, Motilal
Nehru National Institute of Technology
Allahabad, Prayagraj, India

Prof. Haranath Kar

Preface

Computational intelligence mainly includes the techniques that are iterative in nature. These techniques may help in optimization of solution or in learning/training for a prediction and classification of a model. Since last decade, applications of computational intelligence techniques are providing promising results in engineering and science problems. Among these engineering and science fields, biomedical signal processing is one of the important field, which is currently playing a vital role in healthcare industry.

Focusing on the contribution of fields like computational intelligence and soft computing in biomedical signal processing, this book explore the roots, where these fields are joined with each other. For this, basics and fundamental techniques like swarm optimization, prediction, and classification are explained with minimum necessary theory and easy implementation in MatLab. Besides this, the fundamentals of digital signal processing and biomedical signals are also explained in a simple manner. Real dataset of biomedical signals are used to provide real experience of computational intelligence techniques over these datasets. Complete studies are designed with the explanation of problems targeted in these studies. After applying and executing computational intelligence techniques, performance measures are also evaluated to check their performance.

This book will help readers working in the field of computational intelligence and biomedical signal processing. Readers of computational intelligence field will get the idea and basic implementation for applications of their expertise in biomedical signal processing field. On the other hand, readers of biomedical signal processing field will get the familiarity from the topics of computational intelligence field and will be able to formulate problems to solve through computational intelligence techniques.

Bhopal, Madhya Pradesh, India   Mitul Kumar Ahirwal
Jabalpur, Madhya Pradesh, India   Anil Kumar
Roorkee, Uttarakhand, India   Girish Kumar Singh

Acknowledgments

First of all we thank our parents for all things that we have today. We would like to thank our supervisors. We also express thanks to our family members for their support and sacrifices. Special thanks to Dr. N. S. Raghuvanshi (Director, MANIT Bhopal) and Dr. Nilay Kahre (Head of Computer Science and Engineering Department, MANIT Bhopal) for their encouragement and support. Thanks to all the authors and institutes for uploading their datasets for public usage in research and academics. At last but not least, we thank the almighty god for giving us strength to complete this book as a mission.

Contents

List of Figures

List of Tables

About the Authors

M. K. Ahirwal received his B.E. in Computer Science and Engineering from SATI, Vidisha (MP) (affiliated to RGPV, Bhopal, India) in 2009 and completed his M. Tech. in Computer Technology in The National Institute of Technology, Raipur, India in 2011. He completed his Ph.D. in Computer Science and Engineering Department, PDPM IIITDM, Jabalpur, India. Presently, he is an assistant professor in the Department of Computer Science and Engineering, Maulana Azad National Institute of Technology, Bhopal, India. He has several national and international publications in his credit. He is also working on several research projects funded by different agencies. His research area is biomedical signal processing, swarm optimization, brain–computer interface, and healthcare system. He has been involved as a reviewer in various reputed journals.

A. Kumar has received his B.E. from Army Institute of Technology (AIT) Pune, Pune University in Electronic and Telecommunication Engineering, and M.Tech. and Ph.D. degrees from Indian Institute of Technology Roorkee, India, in 2002, 2006, and 2010, respectively. After doctoral work, he has joined as an assistant professor in the Electronic and Communication Engineering Department, Indian Institute of Information Technology Design and Manufacturing, Jabalpur, India from 2009 to July 2016. His academic and research interest is design of digital filters and multirate filter bank, multirate signal processing, biomedical signal processing, image processing, and speech processing.

G. K. Singh received his B.Tech. degree from G.B. Pant University of Agriculture and Technology, Pantnagar, India, in 1981, and Ph.D. degree from Banaras Hindu University, Varanasi, India, in 1991, both in Electrical Engineering. He worked in the industry for nearly five and a half years. Currently, he is a professor in the Electrical Engineering Department, IIT Roorkee, India. He has coordinated a number of research projects sponsored by the CSIR and UGC, Government of India. He is Associate Editor in *IET Renewable Power Generation Journal*. He has also served as Visiting Associate Professor, Pohang University of Science and Technology, Pohang, South Korea, Visiting Professor, Middle East Technical University, Ankara, Turkey, and Visiting Professor, University of Malaya, Kuala Lumpur, Malaysia. His academic and research interest is design and analysis of electrical machines, biomedical signal processing, and renewable energy and power quality. **Dr. Singh secured rank 1 in India and 250 world ranking (top 0.15%) in the subject area "Networking and Telecommunications" as per the independent study done and published by Stanford University, USA, in 2020.**

Chapter 1
Biomedical Signals

1.1 Introduction

In this chapter, the basic fundamental of biomedical signals has been discussed for easy understanding of later chapters. These biomedical signals are the signals that are generated from the human body. Different phenomena and processes occurring inside the human body can be explored and analyzed with the help of these biomedical signals. More specifically, these are also called as bioelectric potentials, i.e., electrical potentials generated by the human body. These signals are generated from every part of the body. Signals from one body part may differ from signals of other body parts. Basically, their basic properties, such as amplitude, frequency, and phase are different from each other. A very simple example to understand these properties is given in later section of the chapter. Biomedical signal processing and analysis area includes all the steps involved in the acquisition/recording of bioelectric potentials from human bodies, for the calculation and observation of intermediate and concluding results that provide some meaningful information about human body.

Some of the very famous biomedical signals are electromyogram (EMG), electroencephalogram (EEG), and electrocardiogram (ECG). EMG signals are the electrical potential of muscle activity that are recorded from the skin surface above the muscle, and the process is known as electromyography. EEG signals are the electrical potential recorded from skull surface over the human brain; they reflect the activities of brain for different tasks performed by humans. The process is known as electroencephalography. ECG signals are electrical activity of the heart recorded from chest part over the heart; they reflect working steps of the heart. The process is known as electrocardiography.

Many other biomedical signals are discovered, and their recording techniques are also invented with the development of electronics equipment and signal processing techniques. Signal recording/acquisition to store them digitally, filtering to reduce

© The Author(s) 2021
M. K. Ahirwal et al., *Computational Intelligence and Biomedical Signal Processing*,
SpringerBriefs in Electrical and Computer Engineering,
https://doi.org/10.1007/978-3-030-67098-6_1

noise, feature extraction for the extraction of meaningful information by processing signal with specific technique, and classification for identification and categorization of several classes as per their properties are fundamental and necessary steps in typical biomedical signal processing study.

In later sections, the basics of digital signals are provided, because understanding of digital signal is a must to understand the biomedical signals. Further ECG and EEG signals are explained in details.

1.2   Basics of Digital Signal Processing

1.2.1   Signals

Signals can be treated as a unique type of data. Generally, signals are originated or generated in the form of sensory data acquired from real world such as seismic vibrations, speech/voice waves, visual images, bioelectric potentials, etc. Digital signal processing (DSP) is the subject that includes concepts, algorithms, mathematics, or some techniques for the analysis and manipulation of digital signals. These digital signals are the digital or discrete version of analog (continuous) signals. The application areas of signal processing are very wide as shown in Fig. 1.1. It includes many areas of science and engineering [1].

General real-world signals are analog by nature. Some examples are light intensity that changes with distance, change of voltage with time, rate of chemical reaction, which depends upon temperature, etc.

A sine wave is shown in Fig. 1.2. With the help of this plot, an easy visualization of terms like amplitude, frequency, and phase is presented.

In this plot, Y axis shows that the amplitude (A) values range from -1 to $+1$, and X axis shows time in seconds. As depicted in the plot, in a duration of 1 s, the wave completes two cycles: one cycle in 0–0.5 s and the second cycle in 0.5–1 s. This means the frequency (f) of this sine wave is 2 Hz. The duration of one cycle is $1/f$ ($1/2 = 0.5$ s). Phase can be explained as the relative displacement of two waves, which has the same frequency. It is also called as *phase difference* or *phase angle*, a complete cycle is referred as 360° of phase angle, as shown in Fig. 1.3. In Fig. 1.3a, a simple case of sine wave with 0° phase angle is shown. The sine wave with 90° phase shift is shown in Fig. 1.3b. For more clear understanding, both the waves are plotted in Fig. 1.3c. It can be noticed that their peak values are at different position with respect to the phase angle scale (horizontal axis) clearly depicting the phase difference of 90°.

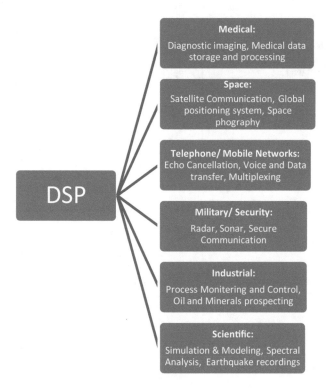

Fig. 1.1 Application areas of digital signal processing

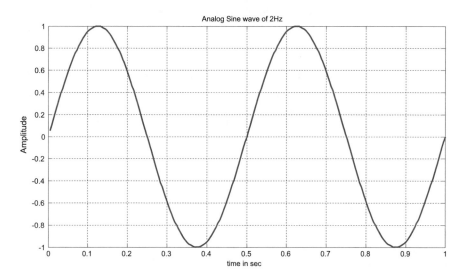

Fig. 1.2 Plot of sine wave

Fig. 1.3 Two sine waves
with 90° phase difference.
(**a**) Sine wave of 1 Hz. (**b**)
Sine wave of 1 Hz with 90°
phase shift. (**c**) Two sine
waves having 90° phase
difference

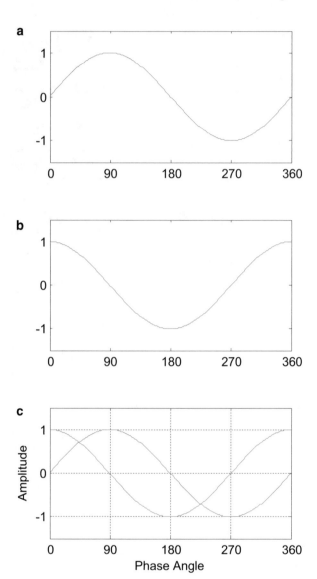

1.2.2 Digital Signals

Continuous signals are referred as analog signals and need to be converted into
digital or discrete signal, then only they are processed through DSP systems. For
this, analog-to-digital conversion (ADC) system is used, which basically works in
two steps, sampling and quantization. These digital signals and the information they
carry may differ from its original continuous signal. This difference is due to

sampling and quantization process. Both these processes decide the amount of information in the digital version of the signal. The process of producing digital signal from analog signal is performed in two steps: the first step is sample-and-hold (S/H) process, and the second step is analog-to-digital converter (ADC). The S/H part will perform sampling; it means accessing the analog signal changes after fixed interval of time. The changes of analog input signal between these sampling points or time are discarded; this is done by keeping (holding) the same voltage value of current sample of analog signal up to next sample. This will generate the flat region of voltage between two samples. Sampling converts the independent variable (time) from continuous to discrete. Now, ADC generates an integer value for each of the flat regions. This produces error, due to loss of some portion of actual voltage values of analog signals. This is known as quantization, i.e., converting the dependent variable (voltage) from continuous to discrete values [1].

Now these aforementioned two concepts are controlled by sampling frequency that decides the time interval between two samples and quantization level that decides the representation of real values to integer values. The quantization depends on how many bits are used to represent the quantized values. It is very simple like with 8-bits, 28 quantization levels are handled, and with 16-bits, 216 quantization levels are handled. On the other hand, sampling frequency decides the number of samples collected in 1 s. If 10 samples are collected in 1 s, then sampling frequency (fs) is 10 Hz, and sampling interval (T) is 1/10 s ($T = 1/fs$).

Then comes the question, how to decide this sampling frequency for any analog or continuous signal? The answer is "sampling theorem"; the definition of this theorem is very simple. Suppose a continuous signal is to be converted into digital signal and then to reconstruct it back to exact analog signal. In that case, first, it is required to choose sampling frequency according to sampling theorem, i.e., "A continuous signal can be properly sampled, only if it does not contain frequency components above one-half of the sampling rate." In a simple manner, the definition is that the sampling frequency must be double the highest frequency present in analog or continuous signal. For example, a sampling rate of 2000 samples/s is required to sample the analog signal of frequency below 1000 cycles/s. In Fig. 1.4, a very simple example is illustrated to visualize the process of sampling and its effect. A sine wave of 2 Hz is plotted, and its two sampled versions are also shown. In the first, the sampling frequency is 100 Hz, and in the second, the sampling frequency is 20 Hz. Now, it is very clear that smaller sampling frequency means less data samples, and high sampling frequency means more data samples.

1.2.3 Digital Filters

Digital filters are mainly used for two purposes: first is separation of mixed or combined signals and second for restoration of signals, which get distorted. Digital filters are more advantageous than analog filters. Analog filters are also termed as electronic filters. Digital filter is a set of coefficients that are calculated by a filter

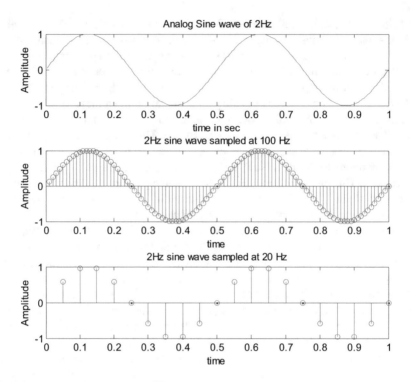

Fig. 1.4 Sampling of sine wave at different sampling frequencies

design algorithm according to desired specification. Many filter design algorithms are available and have their own advantages and disadvantages. Signal separation is required, when a signal gets corrupted due to interference or noise. These are nothing, but other signals that are generated by some unwanted sources or surroundings. For example, consider a machine to measure ECG (electrocardiogram—electrical activity of heart) is placed near some other electrical equipment, and the wires of ECG are in contact with it. Due to contacts of ECG wires with other electrical equipment, unwanted voltage fluctuation distorts the ECG signal. In that case, a digital filter is to be designed according to the characteristics of noise signal to remove it from the acquired ECG signal. Low-pass filter, high-pass filter, and band-pass filter are three common types of filters that are used for this purpose. Low-pass filter is used to block/remove higher frequencies than the specified cutoff frequency. High-pass filter is applied to block/remove low frequencies than the specified cutoff frequency. Band-pass filter blocks the lower and higher frequencies than the specified frequency band. More discussion on this topic is given in Chap. 2.

1.3   Electrocardiogram (ECG) Signal

Heart-related diseases are now a common concern, because of a large number of related diseases. A test through ECG signal helps to identify the ionic activities of heart (cardiac muscles) by placing electrodes (sensors) at different locations on human body over the heart. At present, knowledge of ECG signal analysis is essential for a physician or a cardiologist, to correlate it with the symptoms, and can suggest/recommend the specific treatment. The common structure or pattern of an ECG cycle includes P wave, QRS complex wave, and T wave. A general illustration of ECG cycle is shown in Fig. 1.5a, and real ECG signal is shown in Fig. 1.5b [2, 3]. The standard time for ECG cycle is generally near 0.8 s.

1.3.1   ECG Terminology and Recording

The monitoring of heart through ECG can be done by different instrument/machine using 3, 6, or 12 leads. The standard 12-lead instrument works on 10 electrodes placed over different locations on body as shown in Fig. 1.6 [2, 4]. Figure 1.7 shows the different parts of heart and their contribution in ECG generation. The sinoatrial (SA) node in heart is responsible for ionic activity. Depolarization starts from SA node and spreads throughout the atrium. The depolarization of atria generates P wave for the duration of about 80 ms. The atrioventricular (AV) node spreads the depolarization down to ventricles. The depolarization of left and right ventricles generates QRS complex for a duration of about 80–100 ms. The atrial re-polarization is generally overlapped with the prominent QRS complex, therefore remains hidden in QRS complex. T wave is produced due to re-polarization of ventricles for a duration of around 160 ms. The U wave is generated due to re-polarization of papillary muscle, but is generally ignored [4, 5]. Several intervals in ECG signal are of interest for different studies. Intervals such as PR, ST, and QT interval are also shown in Fig. 1.5.

One complete cycle of ECG signal that includes all the above waves represents a beat. Based on these beats, the heart rate is calculated. Generally, the heart rate of a healthy person lies between 60 and 100 beats per minute in normal conditions [4, 5]. The rhythm of heart is controlled by the impulses produced by the SA node. Disturbance of any type in normal rhythm is called as arrhythmia. Different types of heart arrhythmia are classified with reference to heart rate. If the heart rate is below 60 beats/min, then it is called as bradycardia arrhythmia. In case when the heart rate goes above 100 beats/min, then it is called as tachycardiac arrhythmia [4, 5]. The arrhythmia simply indicates the abnormal working of heart.

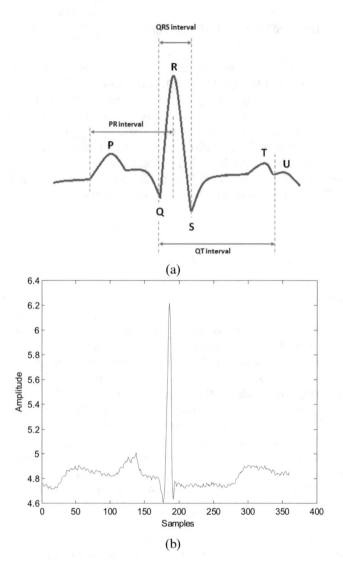

(a)

(b)

Fig. 1.5 (a) ECG cycle with all types of waves and intervals and (b) real ECG signal

1.3.2   ECG Processing

ECG signal contains different and important information about the heart. To extract the information for further analysis, different signal processing methods are used. The ECG signal processing system includes different stages, like pre-processing or filtering, heart rate variability (HRV) analysis, and sometimes compression of ECG [6, 7]. Since, during recording of ECG of any patient, the signal gets attenuated due to the noise introduced by other electromagnetic equipment. Therefore, filtering is

Fig. 1.6 Locations of ECG leads

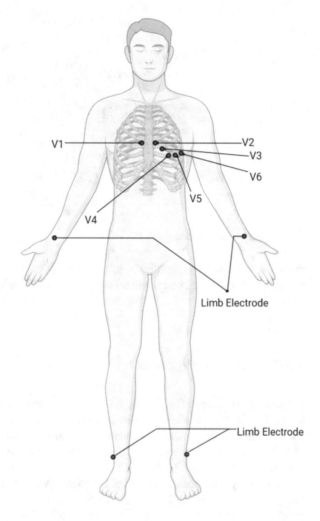

done to remove undesired noise before performing any other signal processing technique. Noise may be of any type, low or high frequencies. So, generally a band pass filter is used. After this, HRV analysis is performed to diagnose the condition of heart and other related diseases.

As the rate of heart beat is a physiological phenomenon associated with the difference in QRS segment duration, HRV analysis includes QRS detection, arrhythmia, diagnosing different intervals of ECG waveform (mainly R-R interval). There is a vital significance of determining the heart rate variability in case of patients suffering from cardiac and mental problems. Bradycardia arrhythmia represents long R-R interval, and tachycardiac arrhythmia represents short R-R interval due to the low and high blood pressure [6]. Thus, HRV analysis is important in diagnosing the disease and is widely accepted by the practitioners. In case of

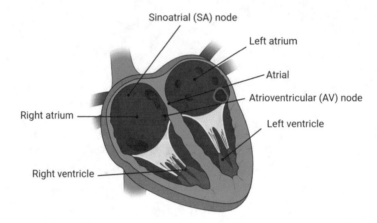

Fig. 1.7 Human heart anatomy and ECG cycle

detection of clinically significant information from ECG, sometimes long ECG recordings need to be stored, and so compression is required to reduce the storage in clinical database.

1.3.3   Common Problems

In ECG signal, different artifacts/noises are introduced, by which information inside the ECG gets affected. Three types of noises that are most common are named as baseline wander, power line interference, and muscle contraction. The visualizations of ECG and artifacts are shown in Fig. 1.8. ECG pre-processing is required to filter out these artifacts [8–10]. Pure ECG is shown in Fig. 1.8a. Baseline wander is a low-frequency noise introduced in ECG which can interfere with the signal, and the noisy ECG signal looks like as shown in Fig. 1.8b. Muscle noise creates major problem in many ECG applications, noisy ECG due to this noise is shown in Fig. 1.8c. Muscle contraction creates electrode motion noises, caused due to the stretching of skin which produces difference in the resistance of skin in contact with the electrode. The resultant ECG due to this is shown in Fig. 1.8d. Therefore, design of filters plays an important role in reducing the artifacts introduced in the signal to be examined.

Sometimes, power line interference is generally introduced due to inappropriate grounding of ECG equipment and also because of the noise and electromagnetic interference from nearby instruments.

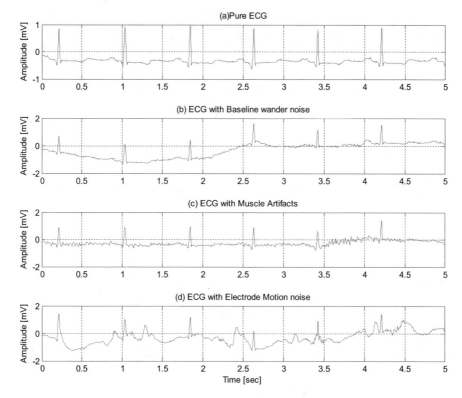

Fig. 1.8 Plots of pure ECG and noisy ECG due to different types of noises. (**a**) Pure ECG. (**b**) ECG with baseline Wander noise. (**c**) ECG with muscle artifacts. (**d**) ECG with electrode motion noise

1.3.4 ECG Applications

- Most commonly used for diagnosing the abnormal functioning of heart.
- Sometimes used for monitoring of a patient under stress.
- Used in general checkups, if the patient is above 40 years.
- In case of chest pain, ECG is done to know about the condition of heart.
- Sometimes, fetal ECG is also done to check the fetal stages.

1.4 Electroencephalogram (EEG) Signal

Human brain is very powerful and the most complicated organ of human body. The working process of brain is not fully explored till now. To explore and understand the working of brain, the easiest and cheapest way is recording of brain waves called as EEG signals. Other techniques like computer tomography (CT) scan and functional magnetic resonance imaging (FMRI) are also available, but they are costly.

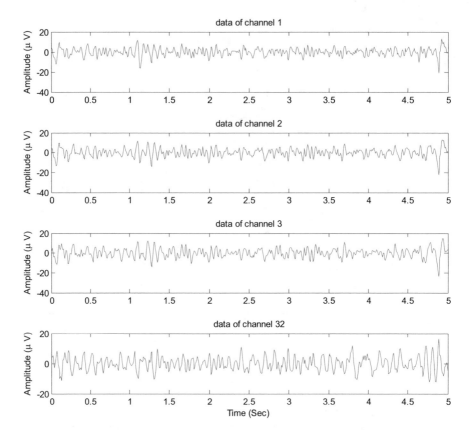

Fig. 1.9 Multi-channel EEG signal

EEG signals are nonstationary and nonlinear by nature, because their generation is complex [11–13]. Combined or collective neural information transferred from one group of neurons to another generates electric potential. This potential is recorded over scalp through noninvasive EEG recording technique [14]. Figure 1.9 shows a multichannel EEG signals. Totally 32-channel EEG data is taken, out of which, EEG signal of channels 1, 2, 3, and 32 is plotted.

Event-related potential (ERP) is an important category of EEG signal. ERP is a specific wave pattern or response of the brain corresponding to any event or task performed by the subject (person). ERP apparels for very short duration of time. ERP characteristic and pattern vary from task-to-task and also depend upon the subject (person whose EEG is recorded) [15, 16]. Figure 1.10 illustrates the ERP signals. Figure 1.10a is a visualization of ERP having P100 wave. This means a positive peak after 100 ms of event or stimulation [17]. Figure 1.10b shows the real visual evoked potential (VEP) recorded from the left occipital (O1) electrode [18]. This pattern was obtained by an average of 30 trials of same events. These ERP signals play an important role in brain–computer interface (BCI) research. These ERP may be time

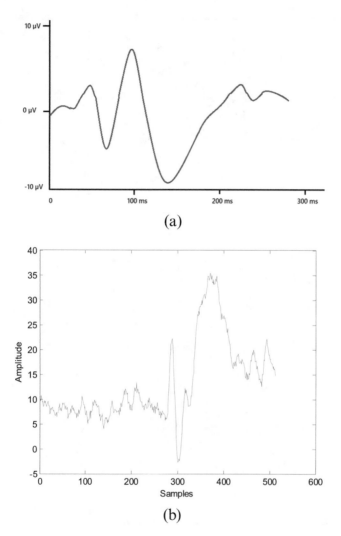

Fig. 1.10 (**a**) ERP with P100 wave and (**b**) real visual evoked potential (VEP)

locked (response of event must occur at fixed time) or non-time locked (response occurs after a variable time). Sometimes, they may also be characterized by phase lock (only if it takes the same phase angle).

1.4.1   Basic Terminology and Recording

In a very basic manner, EEG potential termed as $V(x, y, t)$ is explained as the voltage reading taken at time t from an electrode placed at locations (x, y) over the scalp.

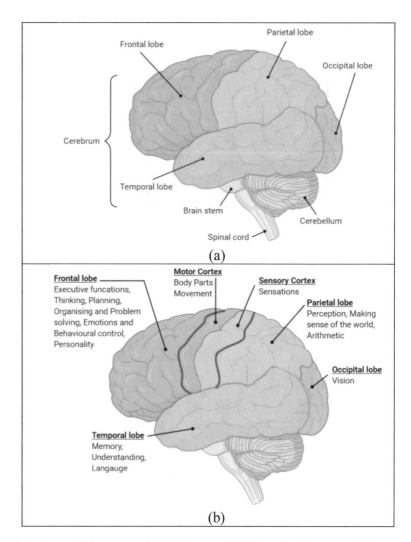

Fig. 1.11 Human brain anatomy: (**a**) brain lobes and (**b**) lobes with their responsibilities

Human brain is divided into left and right hemispheres and further subdivided into different lobes as shown in Fig. 1.11a. Frontal lobe, parietal lobe, occipital lobe, and temporal lobe are responsible for various tasks listed in Fig. 1.11b.

The outer layer of the cerebellum is called as cerebrum cortex. It is a folded structure of 2–5 mm varying thickness. Its total surface area is about 2000 cm^2 and consists of near about 10^{10} (approximately 10 billions) neurons. For recording of EEG, generally a 10–20 electrode placement system is used. This is an international standard of EEG electrode placement and also has some variants, based on the number of electrodes used. In 10–20 electrode placement system, 21–64 channels (electrodes) are used for EEG recording. Sometimes, 10–10 electrode placement system is used with 128 electrodes. Increase in number of electrodes provides high spatial resolution [11, 12]. In Fig. 1.12, an easy visualization of 10–20 electrode placement system is provided.

Electrode placements are labeled as per the names of lobes of brain areas as depicted in Fig. 1.12b, c, like F (frontal lobe), C (central lobe), T (temporal lode), P (Parietal lode), O (occipital lode), and A (earlobes for reference electrodes). The letters are also associated with odd and even numbers at left and right side of the head, respectively. Commonly used sampling frequency for recoding are 256–512 Hz, it can vary up to 20 kHz for research purpose. Different referencing styles are also used in EEG recordings. Some of them are:

Sequential Montage: In this, each channel is represented as the difference of two adjacent electrodes readings. Let channel "Fp1-F3" is the difference of voltages recorded from Fp1 electrode and the F3 electrode.

Referential Montage: In this, each channel is represented as difference of readings of certain electrode and an electrode designated as reference electrode. As such no standard position is fixed for reference. Generally, ear lobe position is used for reference electrode.

Average Reference Montage: The recordings of all electrodes are summed and averaged, now the averaged signal is used as the reference for all channel.

Laplacian Montage: In this, each channel is represented as difference between an electrode and a weighted average of the surrounding electrodes.

Sometimes for clinical purpose and critical cases, EEG signals are also recorded from inside the scalp by surgical operation. This is called invasive EEG recording. The normal recording over the scalp is known as noninvasive EEG recording. The advantage of invasive EEG recording over noninvasive recording is higher amplitudes and spatial resolution.

1.4.2 EEG Processing

Basic steps in EEG signal processing are filtering and analysis. Filtering basically serves two purposes: the first is to remove noise and the second is to separate different EEG frequency bands. Analysis generally aims for feature extraction and classification purposes [12].

EEG frequency spectrum is divided into its conventional frequency bands with their associated mental states. Generally, EEG frequency spectrum is from 0.5 to 35 Hz (full band) with around 100 μV amplitude. Following are the different EEG frequency bands:

1. Delta (δ) waves (frequency band below 4 Hz)—deep sleep
2. Theta (θ) waves (frequency band of 4–8 Hz)—meditation
3. Alpha (α) waves (frequency band of 8–13 Hz)—relaxed state

Fig. 1.12 General visualization of EEG electrode over the scalp, (**a**) General view, (**b**) Lateral view and (**c**) Top view of positions of electrodes in 10–20 electrode placement system

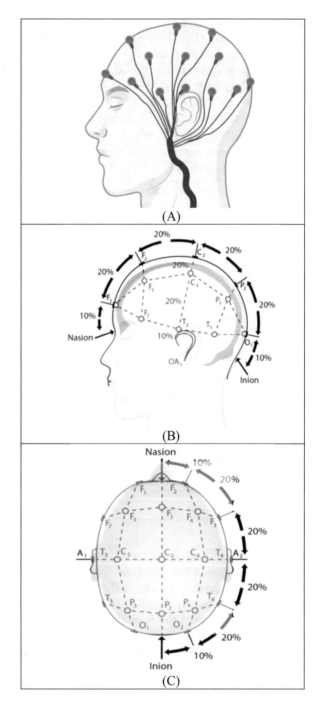

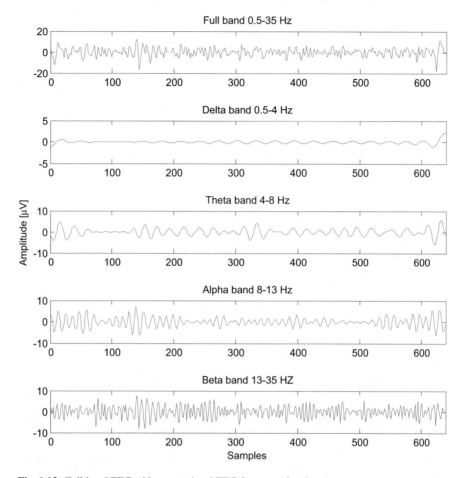

Fig. 1.13 Full-band EEG with conventional EEG frequency bands

4. Beta (β) waves (frequency band of 13–30 Hz)—active or working
5. Gamma (γ) waves (above 30 Hz frequency band)—higher level meditation and hyper activity of brain

Figure 1.13 illustrates the full-band EEG and other EEG bands.

1.4.3   Common Problems

Generally, EEG signal gets corrupted by two types of artifacts, biological and environmental. Biological artifacts are other bioelectric potentials, like electrooculogram (EOG) signal generated due to eye blinks, eye movements, and extra-ocular muscle activity, ECG artifacts, and electromyogram (EMG) signal due to muscle

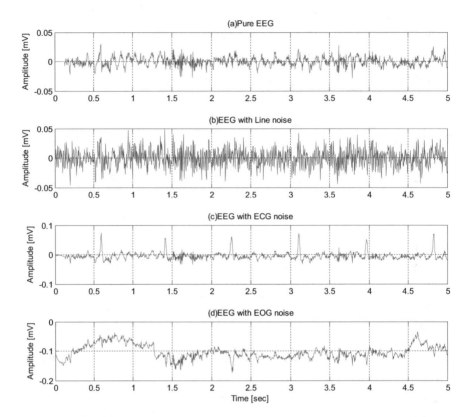

Fig. 1.14 Plots of pure EEG and noisy EEG due to different types of noises. (**a**) Pure EEG. (**b**) EEG with line noise. (**c**) EEG with ECG noise. (**d**) EEG with EOG noise

activity. While environmental artifacts are AC power line noise, electromagnetic field of other devices, electrode cable movements, too much electrode paste/jelly, or dried electrodes. These artifacts can be easily filtered through specific types of filters [19–21]. In Fig. 1.14, a pure EEG and noisy EEG with line noise, ECG and EOG is visualized. Figure 1.14a depicts a pure EEG, in Fig. 1.14b noisy EEG due to line noise is shown. Figure 1.14c, d are of noisy EEG due to ECG and EOG signals.

On the other hand, EEG self-interference is a serious problem in some studies. Since EEG signals are recorded from entire brain, the electrical potential at one location can easily create interference to other nearby locations. This problem is generally solved by blind source separation method. Because, the accurate source of EEG signal inside brain is not known, and it is a collective potential of group of neurons.

1.4.4   EEG Applications

- Clinical diagnosis and analysis for brain disorders
- Used in mental state identification
- Used for emotion, stress, mental workload quantification
- Used in brain–computer interface (BCI) for patients and for entertainment

References

1. S.W. Smith, *The Scientist and Engineer's Guide to Digital Signal Processing* (California Technical Publishing, San Diego, 1997), p. 35. https://www.dspguide.com/
2. https://openstax.org/books/anatomy-and-physiology/pages/19-2-cardiac-muscle-and-electrical-activity
3. P. Pławiak, ECG signals (1000 fragments), Mendeley Data, V3. https://doi.org/10.17632/7dybx7wyfn.3 (2017)
4. J. Malmivuo, R. Plonsey, *Bioelectromagnetism: Principles and Applications of Bioelectric and Biomagnetic Fields* (Oxford University Press, New York, 1995)
5. H. Silva, A. Lourenço, F. Canento, A. Fred, N. Raposo, ECG biometrics: principles and applications, in *Proceedings of the International Conference on Bio-inspired Systems and Signal Processing* (2013), pp. 215–220. https://doi.org/10.5220/0004243202150220
6. O. Navin, G. Kumar, N. Kumar, K. Baderia, R. Kumar, A. Kumar, R-peaks detection using Shannon energy for HRV analysis, in *Advances in Signal Processing and Communication*, Lecture Notes in Electrical Engineering, ed. by B. Rawat, A. Trivedi, S. Manhas, V. Karwal, vol. 526, (Springer, Singapore, 2019). https://doi.org/10.1007/978-981-13-2553-3_39
7. R.S. Anand, V. Kumar, Efficient and reliable detection of QRS-segment in ECG signals, in *Proceedings of the First Regional Conference, IEEE Engineering in Medicine and Biology Society and 14th Conference of the Biomedical Engineering Society of India. An International Meet*, New Delhi, India (1995), pp. 2/56–2/57. https://doi.org/10.1109/RCEMBS.1995.511734
8. M.R. Kose, M.K. Ahirwal, R.R. Janghel, Descendant adaptive filter to remove different noises from ECG signals. Int. J. Biomed. Eng. Technol. **33**(3), 258–273 (2020)
9. G.B. Moody, R.G. Mark, The impact of the MIT-BIH arrhythmia database. IEEE Eng. Med. Biol. **20**(3), 45–50 (2001). PMID: 11446209
10. G.B. Moody, W.E. Muldrow, R.G. Mark, A noise stress test for arrhythmia detectors. Comput. Cardiol. **11**, 381–384 (1984)
11. S. Sanei, J.A. Chambers, *EEG Signal Processing* (Wiley, New York, 2007)
12. S. Tong, N.V. Thakor, *Quantitative EEG Analysis Methods and Clinical Applications* (Artech House Publishers, London, 2009)
13. T.F. Collura, History and evolution of electroencephalographic instruments and techniques. J. Clin. Neurophysiol. **10**(4), 476–504 (1993)
14. S. Koelstra et al., DEAP: a database for emotion analysis; using physiological signals. IEEE Trans. Affect. Comput. **3**(1), 18–31 (2012). https://doi.org/10.1109/T-AFFC.2011.15
15. M. Teplan, Fundamentals of EEG measurement. Meas. Sci. Rev. **2**, 2 (2002)
16. G. Pfurtscheller, F.H.L. Silva, Event-related EEG/MEG synchronization and desynchronization: basic principles. Clin. Neurophysiol. **110**(11), 1842–1857 (1999)
17. J.V. Odom, M. Bach, C. Barber, M. Brigell, M.F. Marmor, A.P. Tormene, G.E. Holder, Visual evoked potentials standard. Doc. Ophthalmol. **108**, 115–123 (2004)

18. R.Q. Quiroga, EEG, ERP and single cell recordings database. http://www.vis.caltech.edu/
    ~rodri/data.htm
19. M.K. Ahirwal, A. Kumar, G.K. Singh, Descendent adaptive noise cancellers to improve SNR of
    contaminated EEG with gradient based and evolutionary approach. Int. J. Biomed. Eng.
    Technol. **13**(1), 49–68 (2013)
20. Y. Ichimaru, G.B. Moody, Development of the polysomnographic database on CD-ROM.
    Psychiatry Clin. Neurosci. **53**(2), 175–177 (1999)
21. A. Goldberger, L. Amaral, L. Glass, J. Hausdorff, P.C. Ivanov, R. Mark, H.E. Stanley,
    PhysioBank, PhysioToolkit, and PhysioNet: components of a new research resource for com-
    plex physiologic signals. Circulation [Online] **101**(23), e215–e220 (2000)

Chapter 2
Fundamentals of Adaptive Filters

2.1 Introduction

Adaptive filters are the filters that are able to change their weights or coefficients by learning algorithm. This means a learning mechanism is the fundamental concept of adaptive filter. In the initial stage of filtering, adaptive filter will learn the characteristics of signals, and accordingly, it modifies its weights. So, after some iteration, adaptive filter starts working fine. The response of filter changes with the signal characters. It has a dynamic kind of nature; initially, weights are set to zero. After some iterations, the weights will be modified by the algorithm used for learning, which is the core part of adaptive filter. A number of algorithms are available which are used inside the adaptive filters. In later sections, some of the algorithms are discussed and demonstrated over biomedical signals.

Before discussing the adaptive filters in detail, some fundamentals about digital signal processing systems are necessary to understand. There are two types of systems: (a) time-invariant and (b) time-variant. Time-invariant systems need all the parameters predefined before the modeling of system, like specification, choice of structure, and choice of algorithm. These systems are of static nature, and once modeled or defined, they remain as it is. While, time-variant systems have the capability to adapt themselves according to input signal. Generally, to start working with these types of systems, only few parameters are sufficient.

Now comes the questions why to use adaptive filters? How to implement them? When to use them? In later sections, answers of all the questions are discussed one-by-one.

© The Author(s) 2021   21
M. K. Ahirwal et al., *Computational Intelligence and Biomedical Signal Processing*,
SpringerBriefs in Electrical and Computer Engineering,
https://doi.org/10.1007/978-3-030-67098-6_2

2.1.1 Why to Use Adaptive Filters

In real world, mostly all real-time systems are surrounded by the environment that is changing with respect to time. Surrounding environment affects the input signal, means that environment is adding noise to input signal, which is being processed by the system. As a result, a system has a surrounding which creates nonstationary or nonlinear inputs. In such cases, static systems are unable to perform well, as they have no mechanism to learn or sense the environment or changes in the input. In adaptive filter, this capability is inherited by the use of learning algorithm. Hence, adaptive filters are more suitable in real-life problems.

Below, some simple examples are discussed, which will clear the problem or situations, in which adaptive filters are required.

In Fig. 2.1, a case of mixture of 4 Hz sine wave and 5–10 Hz noise signal is illustrated. These signals are easily separated by using low pass filter of 5 Hz cutoff frequency, because frequency spectrums of sine wave and noise are not overlapping. In Fig. 2.2, a case of 4–10 Hz sine wave and 3–25 Hz noise signal are illustrated; their frequency bands are overlapped with each other. It is difficult to use low pass or high pass filter in this case. Because, below 5 Hz, there is noise, and so low pass filter cannot be used. Signal of interest will also be filtered, if high pass filter is used. This case can be successfully handled by the adaptive filters [1–4].

2.1.2 Models of Adaptive Filters

In this section, the answer to question how to implement adaptive filter? will be addressed. Figure 2.3 illustrates the model of adaptive filter for the above-discussed problem of overlapped frequency bands. Here, noisy signal (sine wave of

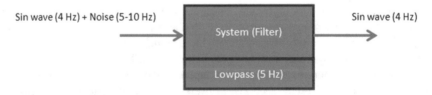

Fig. 2.1 Low pass filter for the removal of noise signal (non-overlapped frequency bands)

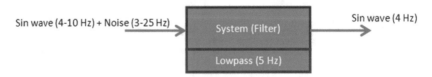

Fig. 2.2 Low pass filter for the removal of noise signal (overlapped frequency bands)

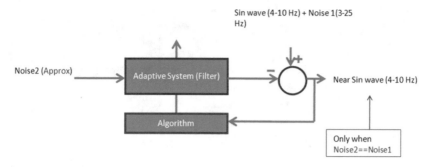

Fig. 2.3 Simple model of adaptive filter as adaptive noise canceller (ANC)

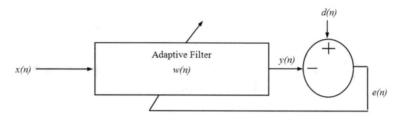

Fig. 2.4 Basic model of an adaptive filter

4–10 Hz + noise 1 signal of 3–25 Hz) and one more noise 2 signal is taken. This noise 2 signal is the approximation of noise 1 signal; this is also known as the reference signal. The target of adaptive filter is to convert this noise 2 signal same as noise 1, so that it can be subtracted from noisy signal to get pure signal (sine wave). This model is generally referred as adaptive noise canceller (ANC). In later section, other models are also discussed.

The basic or generalized model/structure of adaptive filter is illustrated in Fig. 2.4. Signal sample index is represented as n in discrete time domain. The input signal is represented as $x(n)$, and $d(n)$ represents the reference signal, which acts as desired output signal (some noise components are present in it). Output signal is $y(n)$ and error signal is $e(n)$. Error signal is simply obtained as $e(n) = d(n) - y(n)$.

Now, the question when to use adaptive filters? is addressed. In case, when the system specifications are unknown or it is difficult to define them, like unable to define cutoff frequency. For a real-time model in dynamic environment, this means system needs changes or adaptation in real-time. Some cases, where adaptive filter is used, are channel or system identification, eco-cancellation, active noise cancellation, channel equalization, etc.

For different cases or situations, some models of adaptive filter have been developed and used as standard models [1–4]. Basically, there are three models: (a) system identification, (b) channel equalization, and (c) noise canceller. The structures of these models are shown in Fig. 2.5. Noise canceller model is discussed above with an example of sine wave signal.

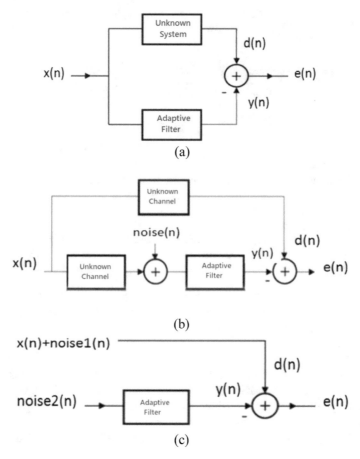

Fig. 2.5 Adaptive filter models as (**a**) system identification, (**b**) channel equalization, and (**c**) noise canceller

2.1.3 Algorithms for Adaptive Filter

As discussed above that the core part of adaptive filter is algorithm, which works inside it. This algorithm is basically used for learning and updating of weights or coefficients of filter. The simplest description of work mechanism of this algorithm is given below and is derived from the gradient-based approach:

$$(\text{updated value of weights}) = (\text{old weights}) + (\text{learning rate}) \times (\text{error}) \times (\text{input signal}).$$

The above mechanism is repeated in every iteration of algorithm. Furthermore, precise terms in this mechanism are weights as $W = [w(0), w(1), w(2), \ldots w(M-1)]$

of size $1 \times M$, constant learning rate as μ, error signal is $e(n)$ and input signal is $x(n)$, where n is current index.

Initial implementation of adaptive filter can be easily done by gradient-based algorithms. These are the best options to explore the working of adaptive filters. In the later section, few gradient-based algorithms are given with their equations and are explained as noise canceller model [2].

2.1.3.1 Least Mean Square (LMS) Algorithm

The LMS algorithm is very famous and easiest algorithm to implement adaptive filter, because of its simple steps and low computational complexity [6, 7, 11]. In this algorithm, estimation of error and updating of weights are done through Eqs. (2.1)–(2.3), respectively.

$$y(n) = \sum_{k=0}^{M} W(n)_k \times x(n - k) \tag{2.1}$$

$$e(n) = d(n) - y(n) \tag{2.2}$$

$$W(n + 1) = W(n) + 2 \times \mu \times e(n) \times x(n) \tag{2.3}$$

In above equation, $x(n)$ is the input vector (signal correlated to noise), and $d(n)$ represents the desired signal. $y(n)$ denotes the estimate of desired signal, and $e(n)$ is the error estimation. $W(n)$ represents the weight vector at nth iteration, and learning rate is denoted by μ, also called as step size. The learning rate regulates the weights convergence.

2.1.3.2 Normalized Least Mean Square (NLMS) Algorithm

To improve the convergence of weights, normalized LMS algorithm is developed; this does not require any more calculation. This is just a variant of LMS algorithm. The NLMS algorithm uses the concept of variable learning rate or convergence factor for minimization of instantaneous output error [1–4]. Additional changes in NLMS algorithm are given in Eq. (2.4),

$$W(n + 1) = W(n) + \frac{\mu}{\kappa + x(n)^T \times x(n)} \times e(n) \times x(n) \tag{2.4}$$

In the above equation, κ represents a constant that holds very small value, and the initial step is μ_n having value in the range of 0–2.

When applying adaptive filter, few parameters and settings need to be fixed like length (M) of the adaptive filter, value of step size (μ), and algorithm (LMS or

NLMS). For providing practical application for these algorithms, MatLab functions are designed, as given in LMS_algorithm.m and NLMS_algorithm.m.

```
LMS_algorithm.m
function[w,y,e]=LMS_algorithm(x,d,mu,M)

% x=input signal
% d= desired or reference signal
% mu=learning rate
% M=order of filter

N=length(x);
y=zeros(1,N);
w=zeros(1,M);

for n=M:N
    x1=x(n:-1:n-M+1);
    y(n)=w*x1';                % output signal eq.(1)
    e(n)=d(n)-y(n);            % error signal eq.(2)
    w=w+2*mu*e(n)*x1;          % weight update eq.(3)
    w1(n-M+1,:)=w(1,:);
end;
```

```
NLMS_algorithm.m
function[w,y,e]=NLMS_algorithm(x,d,mu,M)

% x=input signal
% d= desired or reference signal
% mu=learning rate
% M=order of filter
% k=constant

k=0.001;
N=length(x);
y=zeros(1,N);
w=zeros(1,M);

for n=M:N
    x1=x(n:-1:n-M+1);
    y(n)=w*x1';
    e(n)=d(n)-y(n);
    w=w+(mu/(k+x1*x1'))*e(n)*x1; % weight update
eq.(4)
    w1(n-M+1,:)=w(1,:);
end;
```

For providing practical exposure to adaptive noise canceller, MatLab scripts are provided as MatLab functions, very basic examples are demonstrated below:

2.1.3.3 Example Cases

Example Case 1: Noise removal from sinusoidal signals (sine wave). In this, sine wave of 4 Hz is generated and treated as pure signal $x(n)$. To convert pure signal into noisy signal, noise signal is added. Noise signal is generated through random function with 30% amplitude. Learning rate is fixed to 0.04 and filter length is 5. This setup is provided in script_ch2_1.m.

```
script_ch2_1.m
clc;
clear all;
close all;

D =   4; % signal duration (sec)
fs = 250; % sampling rate (samples per sec)
T = 1/fs; % sampling period
t = [T:T:D]; % time vector

x = sin(2*pi*4*t); %pure signal x(n)
ln = 0.3*randn(size(t));   %noise signal
ns=x+ln;                   %noisy signal d(n)

figure
subplot(311);plot(x);ylabel('x(n)')
subplot(312);plot(ln);ylabel('noise')
subplot(313);plot(ns);ylabel('d(n)')

% Apply ANC
M=5; %filter length
mu=0.04; %learning rate
[w,y,e]=LMS_algorithm(ln,ns,mu,M);

figure
subplot(311);plot(ns);ylabel('d(n)')
subplot(312);plot(e);ylabel('e(n)')
subplot(313);plot(y);ylabel('y(n)')
```

Signals used in this setup are illustrated in Fig. 2.6.

After execution of this setup, observation is easily done by plotting of signals as in Fig. 2.7.

From the plot of error and output signal, it is easily observed that after some iteration, $e(n)$ acquires the shape of pure signal, and the noise signal is accumulated in $y(n)$ as noise. In each iteration, this noise is removed from the desired

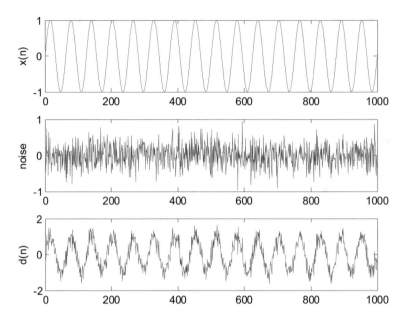

Fig. 2.6 Plot of pure signal $x(n)$, noise signal, and noisy signal $d(n)$

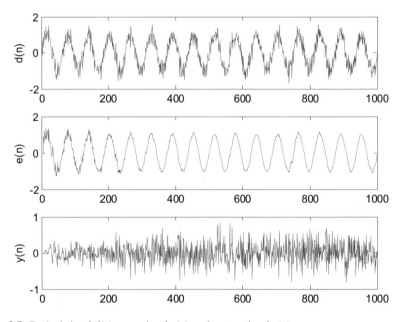

Fig. 2.7 Desired signal $d(n)$, error signal $e(n)$, and output signal $y(n)$

signal *d*(*n*). In this setup of ANC, estimation of pure signal (also called as filtered signal) is accumulated in *e*(*n*).

Example Case 2: Noise removal from sinusoidal signals (sine wave) having multiple frequencies. In this, sine wave as mixture of 10, 20, and 30 Hz is generated and treated as pure signal *x*(*n*). To convert pure signal into noisy signal, noise signal is added. Noise signal is generated through random function with 70% amplitude. Learning rate is fixed to 0.04 and filter length is 5. This setup is provided in script_ch2_2.m.

```
script_ch2_2.m
clc;
clear all;
close all;

D =  1; % signal duration (sec)
fs = 1000; % sampling rate (samples per sec)
T = 1/fs; % sampling period
t = [T:T:D]; % time vector

x = sin(2*pi*10*t)+sin(2*pi*20*t)+sin(2*pi*30*t);
%pure signal x(n)
ln = 0.7*randn(size(t));   %noise signal
ns=x+ln;                   %noisy signal d(n)

figure
subplot(311);plot(x);ylabel('x(n)')
subplot(312);plot(ln);ylabel('noise')
subplot(313);plot(ns);ylabel('d(n)')

% Apply ANC
M=5; %filter length
mu=0.04; %learning rate
[w,y,e]=LMS_algorithm(ln,ns,mu,M);

figure
subplot(311);plot(ns);ylabel('d(n)')
subplot(312);plot(e);ylabel('e(n)')
subplot(313);plot(y);ylabel('y(n)')
```

Signals used in this setup are illustrated in Fig. 2.8.

After execution of this setup for example 2, observation is easily done by plotting of the signals as in Fig. 2.9. Same type of observation is there as previous case.

Example Case 3: Noise removal from sinusoidal signals (sine wave) having multiple frequencies. All the signals are same as previous example. Filter length is changed to 3 and learning rate is taken 0.01. This setup is provided in script_ch2_3.m.

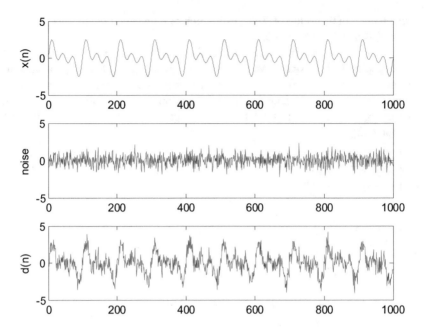

Fig. 2.8 Plot of pure signal $x(n)$ having multiple frequencies, noise signal, and noisy signal $d(n)$

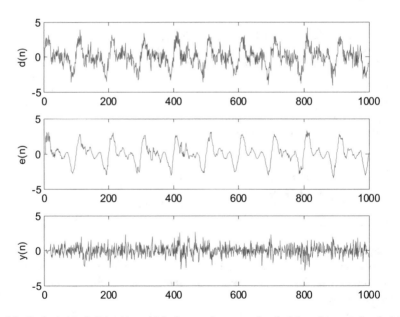

Fig. 2.9 Desired signal $d(n)$ with multiple frequencies, error signal $e(n)$, and output signal $y(n)$

```
script_ch2_3.m
clc;
clear all;
close all;

D =  1; % signal duration (sec)
fs = 1000; % sampling rate (samples per sec)
T = 1/fs; % sampling period
t = [T:T:D]; % time vector

x = sin(2*pi*10*t)+sin(2*pi*20*t)+sin(2*pi*30*t);
%pure signal x(n)
ln = 0.7*randn(size(t));   %noise signal
ns=x+ln;                   %noisy signal d(n)

figure
subplot(311);plot(x);ylabel('x(n)')
subplot(312);plot(ln);ylabel('noise')
subplot(313);plot(ns);ylabel('d(n)')

% Apply ANC
M=3; %filter length
mu=0.01; %learning rate
[w,y,e]=LMS_algorithm(ln,ns,mu,M);

figure
subplot(311);plot(ns);ylabel('d(n)')
subplot(312);plot(e);ylabel('e(n)')
subplot(313);plot(y);ylabel('y(n)')
```

After execution of this setup for example 3, for observation plotting of signals are provided in Fig. 2.10.

In this example 3, some changes are observed as compared to example 2. By changing the filter length and learning rate, the quality of the filtered signal is better than the previous example.

The summary of the above examples is listed in Table 2.1. This table also concludes that the adaptive filters are totally independent on the specification and characteristics of signals. As in example case 1 and 2, the adaptive filter of same length and step size works fine with signals having different frequencies and percentage/level of noise added to them. In case 2 and 3, only by changing the filter length and learning rate, quality of filtered signal gets improved, while input and desired signals remain same.

2.1.3.4 Fidelity Parameters for Observation

To observe the changes in the quality of filtered signals, some fidelity parameters are used. These fidelity parameters will evaluate the signal quality in quantitative

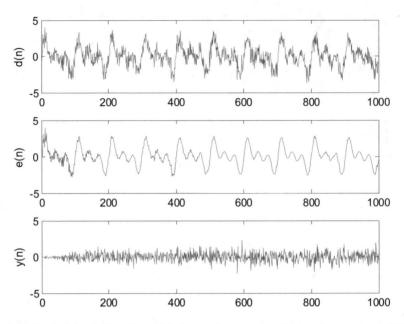

Fig. 2.10 Desired signal $d(n)$ with multiple frequencies, error signal $e(n)$, and output signal $y(n)$ for new filter length and learning rate

Table 2.1 Specifications of signals and adaptive noise cancellers in the above example cases

Example case	Algorithm	Pure signal	Noise signal	Step size (μ)	Length of filter
1	LMS	Sine (4 Hz)	30% random noise	0.04	5
2	LMS	Sine (10 Hz + 20 Hz + 30 Hz)	70% random noise	0.04	5
3	LMS	Sine (10 Hz + 20 Hz + 30 Hz)	70% random noise	0.01	3

manner [5–7]. Some of the famous fidelity parameters are signal-to-noise ratio (SNR dB), mean square error (MSE), and maximum error (ME), explained below in Eqs. (2.5)–(2.8):

- SNR at input:

$$\text{SNR}_{\text{IN}} = 10 \log_{10} \frac{\sum \text{input}^2}{\sum \text{output}^2} \tag{2.5}$$

- SNR at output:

$$\text{SNR}_{\text{OUT}} = 10 \log_{10} \frac{\sum \text{input}^2}{\sum (\text{input} - \text{output})^2} \tag{2.6}$$

- Mean square error (MSE):

$$\text{MSE} = \frac{1}{N} \sum (\text{input} - \text{output})^2 \tag{2.7}$$

- Maximum error (ME):

$$\text{ME} = \max\left(\text{abs}(\text{output} - \text{input})\right) \tag{2.8}$$

The above fidelity parameters can be easily coded as MatLab functions as provided in fidelity_parameters.m.

```
fidelity_parameters.m
function [SNRin SNRout MSE
ME]=fidelity_parameters(in, no, ou)
% in = input or desired signal
% no = noise signal
% ou = output signal
%---------SNRin-----------------
inp=0;
for i=1:length(in)
    inp= inp+ in(i)*in(i);
end;
nop=0;
for i=1:length(no);
    nop=nop+ no(i)*no(i);
end;
SNRin=10*log10(inp/nop);

%---------SNRout----------------
diff=0;
for i=1:length(ou);
    diff=diff+((in(i)-ou(i))*(in(i)-ou(i)));
end;
SNRout=10*log10(inp/diff);
%---------MSE-------------------
MSE=(1/length(ou))*diff;
%----------ME-------------------
di=0;
for i=1:length(ou);
    di(i)=abs(((ou(i)-in(i))));
end;
ME=max(di);
```

Quantitative evaluation of the above example cases is listed in Table 2.2, in the form of the above-discussed fidelity parameters. The values of these fidelity parameters are obtained by using the above function just after the execution of each

Table 2.2 Performance of adaptive filter in example cases 1, 2, and 3

Example case	SNR$_{IN}$	SNR$_{OUT}$	MSE	ME	Remark on SNR
1	7.9202	8.8924	0.0759	1.0367	Increases, means noise removed
2	5.9342	5.0960	0.6214	3.3147	No significant difference
3	5.7817	7.2786	0.3495	2.0340	Increases, means noise removed

example case or call to above function can be appended at the last of each case (script file) as:

```
[SNRin, SNRout, MSE, ME]=fidelity_parameters(ns, ln, e);
```

In the above function call, with reference to Eqs. (2.5)–(2.8) and fidelity_parameter.m function, noisy signal is passed as input, noise signal passed as noise and error signal is passed as output, because in $e(n)$, filtered signal is accumulated.

The results listed in Table 2.2 may change on execution of script, because every time noise signal (random signal) gets changed. As changes observed in Table 2.2, in a similar manner, adaptive filters can easily be applied over different biomedical signals. Some applications of adaptive filter used over electroencephalogram (EEG) and electrocardiogram (ECG) signals are discussed and practically demonstrated.

2.2 Application of Adaptive Filter over EEG Signals

In this section, real-life problems are taken as examples. Dataset used in this section is taken from Physionet website [8]. This dataset consists of EEG signal recordings that are corrupted by ECG signals and line noise. Adaptive filter has been applied over EEG signal, and one-by-one, these noises are removed.

2.2.1 EEG Dataset

The time of recording is 10 s with a sampling frequency of 250 Hz. Complete dataset consists of several types of signals (ECG, EEG, respiration (nasal), respiration (abdominal), electrooculogram (EOG), and electromyogram (EMG)). For this example, EEG, EOG, and ECG signals are selected from the dataset. Line noise was generated artificially as a sine wave having 48, 50, and 52 Hz frequencies with some amount of random noise mixed to it. Records of patient with ID P32, P37, P41, P45, and P48 were selected. Figure 2.11 shows all the signals of any specific patient. In this, peak of ECG signal can be easily seen in EEG signal at the same time interval, also the effect of drift in EOG is also visible in EEG signal.

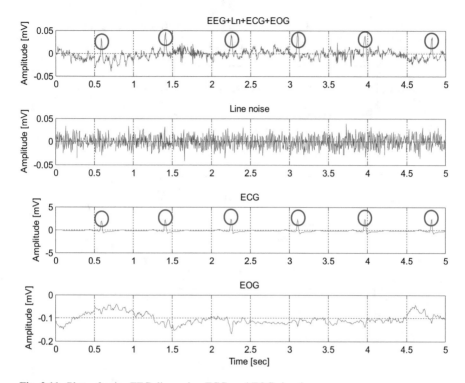

Fig. 2.11 Plots of noisy EEG, line noise, ECG, and EOG signals

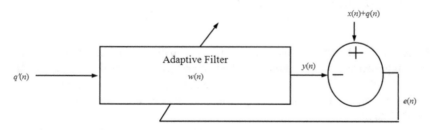

Fig. 2.12 Adaptive noise canceller (ANC) model

2.2.2   Adaptive Filter Model for EEG

Adaptive noise canceller (ANC) model is used to filter the EEG signal, as illustrated in Fig. 2.12. In this model, $x(n)$ is the EEG signal, which gets corrupted by the noise signal $q(n)$. Further in place of this, $q(n)$, line noise, ECG, and EOG signals are taken

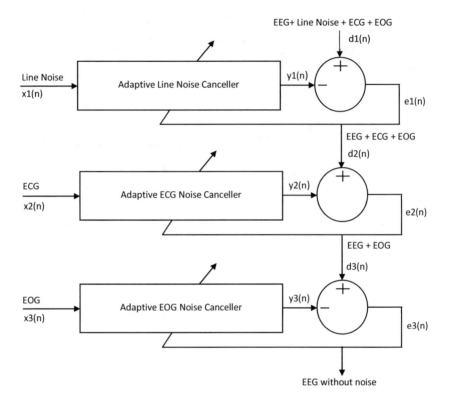

Fig. 2.13 Final descendant model of ANC for EEG

as noises. $q'(n)$ is the correlated version (approximation) of noise. Desired signal d (n) is assumed as a combination of $x(n)$ and noise $q(n)$. This model is repeated three times in a cascade or descendant manner. Each time, different noise is considered as reference signal for filtering [5–7]. The final model implemented is shown in Fig. 2.13 [7].

Now, the above model can be simply implanted as MatLab script using LMS and NLMS adaptive filter algorithm provided as MatLab function. In the same script, fidelity parameters can also be calculated. The ready-to-use script is provided as script_ch2_4_EEG.m.

```
script_ch2_4_EEG.m
% before using data files, please site the data set
as mentioned on below link:
% Physionet-MIT-BIH Polysomnographic
% Database:
https://physionet.org/content/slpdb/1.0.0/
clc;
clear all;
close all;


load p32;   % loading patient record from mat file
p32.mat
% p32 dimension is 4 x 2500
% row 1 is noisy EEG, row 2 is line noise
% row 3 is ECG and row 4 is EOG


val = p32;
fs = 250;   % sampling frequency
D = 10;     % 10 sec data
N = D*fs;   % number of samples
T = 1/fs;   % sampling period
t = [T:T:D]; % time vector


eeg=val(1,:); % separate EEG from data file
ln=val(2,:);
ecg=val(3,:);
eog=val(4,:);
N=length(eeg);
%=========Figure1==================
figure
subplot(4,1,1); plot(t(1:1250),eeg(1:1250));grid;
title('EEG+Ln+ECG+EOG '); ylabel('Amplitude [mV]');
subplot(4,1,2); plot(t(1:1250),ln(1:1250)); grid;
title('Line noise'); ylabel('Amplitude [mV]');
subplot(4,1,3); plot(t(1:1250),ecg(1:1250)); grid;
title('ECG'); ylabel('Amplitude [mV]');
subplot(4,1,4); plot(t(1:1250),eog(1:1250)); grid;
title('EOG'); ylabel('Amplitude [mV]');
xlabel('Time [sec]');
%===========================================

%====== ANC 1 ============================
[w1 y1 e1]=NLMS_algorithm(ln,eeg,0.05,16);
% ln = line noise as reference signal
% eeg = noisy eeg
% 0.05 is step-size
% 16 is filter order
```

```
%====== ANC 2 =======================
[w2 y2 e2]=NLMS_algorithm(ecg,e1,0.02,32);
% ecg= ecg signal as reference signal
% e1 = eeg signal filtered from ANC1
% 0.02 is step-size
% 32 is filter order

%====== ANC 3 =======================
[w3 y3 e3]=NLMS_algorithm(eog,e2,0.02,32);
% eog= eog signal as reference signal
% e2 = eeg signal filtered from ANC2
% 0.02 is step-size
% 32 is filter order

%====== Figure 2 ================

figure
subplot(3,1,1); plot(t(1:1250),e2(1:1250));grid;
title('(EEG - ln - ECG) + EOG'); ylabel('Amplitude
[mV]');
subplot(3,1,2); plot(t(1:1250),ecg(1:1250)); grid;
title('ECG'); ylabel('Amplitude [mV]');
subplot(3,1,3); plot(t(1:1250),y2(1:1250)); grid;
title('error signal adaptation similar to ECG');
ylabel('Amplitude [mV]');


%======== fidelity Parameters======
[SNRin1, SNRout1, MSE1,
ME1]=fidelity_parameters(eeg, ln, e1);
[SNRin2, SNRout2, MSE2, ME2]=fidelity_parameters(e1,
ecg, e2);
[SNRin3, SNRout3, MSE3, ME3]=fidelity_parameters(e2,
eog, e3);

fprintf('SNRin      SNRout      MSE        ME\n');
disp([SNRin1, SNRout1, MSE1, ME1])
disp([SNRin2, SNRout2, MSE2, ME2])
disp([SNRin3, SNRout3, MSE3, ME3])

result=[SNRin1, SNRout1, MSE1, ME1;
        SNRin2, SNRout2, MSE2, ME2;
        SNRin3, SNRout3, MSE3, ME3];
```

2.2.3 Observation

After the execution of the model over EEG data with LMS algorithm, following observation are easily captured through plotting of input and output signals. Figure 2.14 shows how the ECG peaks are removed from the EEG signal. It is also noticed that, with the progress in iterations, error signal takes the shape and characteristics of ECG signal that is removed from EEG signal. The above script also works with NLMS algorithm, just replace the "LMS_algorithm" by "NLMS_algorithm." For quantitative evolution of adaptive filters with different algorithm, fidelity parameters are calculated (as result) and listed in Tables 2.3 and 2.4, for LMS and NLMS algorithms, respectively.

To conclude the overall study, average SNR gain ($SNR_{OUT} - SNR_{IN}$) is calculated for each patient and compared as shown in Fig. 2.15. On an average, both the algorithms are performing same, but their performance differ from patient to patient.

2.3 Application of Adaptive Filter over ECG Signals

In this section, an example of ECG signals is taken. This ECG dataset is taken from Physionet website [9, 10]. This dataset consists of ECG signal recordings, while for noise signals, baseline wonder (BW), musical artifact (MA), and electrode motion (EM) signals are taken from the same website. Adaptive filter is applied over the ECG signals, and one-by-one these noises are removed.

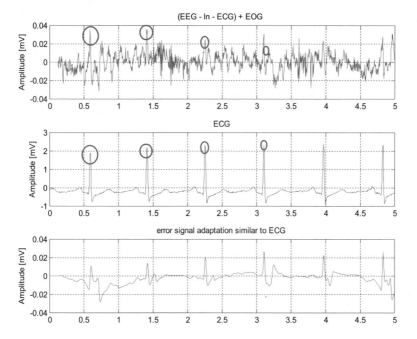

Fig. 2.14 Illustration of filtered signal, ECG, and error signal

Table 2.3 Performance of ANC based on LMS algorithm for EEG filtering

Patient record	ANC	SNR_{IN}	SNR_{OUT}	MSE	ME
P32	ANC1	−0.17576	27.26941	2.68E−07	0.014066
	ANC2	−30.2273	3.488235	6.38E−05	0.045698
	ANC3	−20.4156	17.38976	1.77E−06	0.003573
P37	ANC1	−2.27811	29.21594	1.05E−07	0.007265
	ANC2	−22.4225	8.367815	1.28E−05	0.018953
	ANC3	−23.0777	2.090216	4.58E−05	0.035658
P41	ANC1	2.043785	20.18145	2.28E−06	0.022702
	ANC2	−24.4763	5.60499	6.49E−05	0.043272
	ANC3	−15.6155	8.095316	2.95E−05	0.029155
P45	ANC1	−0.9993	23.46323	5.32E−07	0.018543
	ANC2	−28.8492	4.551977	4.12E−05	0.045389
	ANC3	−9.83911	30.15482	8.39E−08	0.001081
P48	ANC1	0.265667	23.01769	7.89E−07	0.021521
	ANC2	−24.3693	11.05583	1.23E−05	0.043204
	ANC3	−3.81402	43.03903	7.56E−09	0.000525

Table 2.4 Performance of ANC based on NLMS algorithm for EEG filtering

Patient record	ANC	SNR_{IN}	SNR_{OUT}	MSE	ME
P32	ANC1	0.17576	18.46116	2.03E−06	0.014066
	ANC2	−30.1241	5.147722	4.46E−05	0.032006
	ANC3	19.5309	10.74233	1.00E−05	0.008899
P37	ANC1	−2.27811	20.87268	7.20E−07	0.007265
	ANC2	−22.3084	5.655474	2.45E−05	0.02524
	ANC3	−22.4908	−0.00101	8.49E−05	0.048738
P41	ANC1	2.043785	17.56236	4.17E−06	0.022702
	ANC2	−24.3523	5.943272	6.17E−05	0.042916
	ANC3	−14.6553	4.576598	8.28E−05	0.04319
P45	ANC1	−0.9993	17.22265	2.24E−06	0.018543
	ANC2	−28.7667	6.604437	2.62E−05	0.031958
	ANC3	−9.14023	10.27444	9.58E−06	0.013376
P48	ANC1	0.265667	17.75244	2.65E−06	0.021521
	ANC2	−24.2699	8.440583	2.30E−05	0.043027
	ANC3	−3.3756	17.34664	3.10E−06	0.006923

2.3.1   ECG Dataset

This dataset contains 48.5 h records of two leads ECG signal at 360 Hz sampling frequency. Only five subjects recording of 10 s have been taken. Noise signals considered in this dataset are baseline wander noise, muscle artifact noise, and electrode motion noise. To perform the simulation of filtering process, noise is added to ECG signal. First of all, noise signals are mixed together, as illustrated in

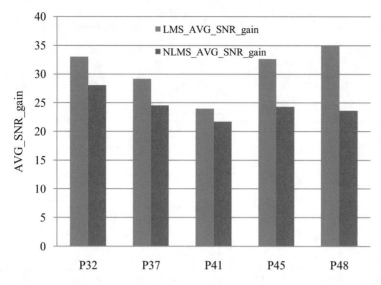

Fig. 2.15 Patient-wise comparison of LMS and NLMS-based ANC

Fig. 2.16. Then, this mixed noise is added with pure ECG at noise level of −10 and −8 dB.

To add noise into the signal, different methods can be used. A simple implementation to add noise in signal is given as MatLab function in add_noise.m.

```
add_noise.m
function [noisy_signal,new_SNR] = add_noise(signal,
SNR_dB, noise)

N = length(noise); % Number of input time samples

signal_power = sum(abs(signal).*abs(signal))/N;

noise_power = sum(abs(noise).*abs(noise))/N;

ratio = (signal_power/noise_power)*10^(-SNR_dB/10);

new_noise = sqrt(ratio)*noise; % Change Noise level

new_noise_power =
sum(abs(new_noise).*abs(new_noise))/N;

new_SNR = 10*(log10(signal_power/new_noise_power));

noisy_signal = signal + new_noise;
```

Script for adding noise in pure ECG is given with complete setup script for this study. In Fig. 2.17, pure and noisy ECG signals are illustrated.

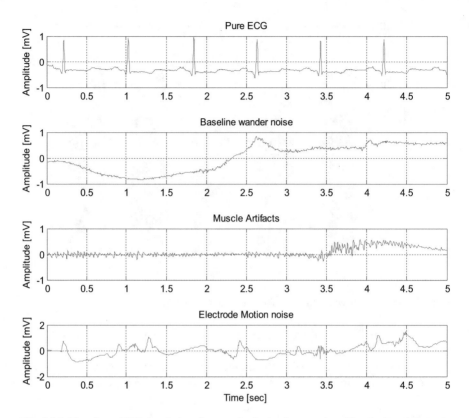

Fig. 2.16 Plot of pure ECG signals, baseline wander (bw) noise, muscle artifact (ma), and electrode motion (em) artifact

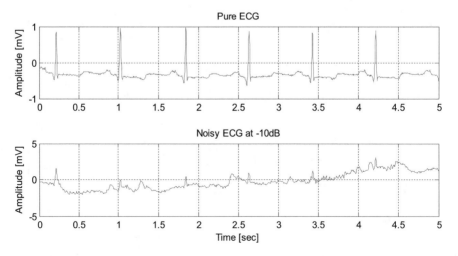

Fig. 2.17 Pure ECG and noisy ECG

2.3.2 Adaptive Filter Model for ECG

Adaptive noise canceller model is used to filter the ECG signal, as illustrated in Fig. 2.12. In this model, $x(n)$ is the ECG signal, which gets corrupted by the noise signal $q(n)$. Further in place of this $q(n)$, BW, MA, and EM signals are taken as noises. $q'(n)$ is the correlated version (approximation) of noise. Desired signal $d(n)$ is assumed as a mixture of $x(n)$ and noise $q(n)$. This model is repeated three times in a cascade manner, each time different noise is considered as reference signal for filtering [11]. The final model implemented is shown in Fig. 2.18.

Now, the above model can be implanted in the same manner as discussed in the previous example. MatLab script with LMS and NLMS adaptive filter algorithm function is used. In the same script, fidelity parameters can also be calculated. The ready-to-use script is provided as script_ch2_5_ECG.m.

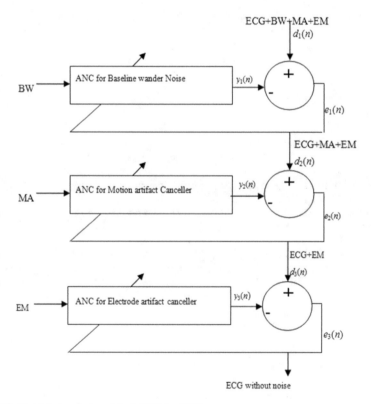

Fig. 2.18 Final descendant model of ANC for ECG

script_ch2_5_ECG.m

```
% before using data files, please site the data set as
mentioned on below link:
% Physionet-MIT-BIH Arrhythmia Database and MIT-BIH
Noise Stress Test Database
% Database: https://physionet.org/content/mitdb/1.0.0/
% Database: https://physionet.org/content/nstdb/1.0.0/
clc;
clear all;
close all;


load ml100;  % loading patient record from mat file
ml109.mat
load noise_bw; % loading Baseline Wander noise
load noise_ma; % loading muscle artifacts
load noise_em; % loading electrode motion noise


% ml100 dimension is 1 x 3600


fs = 360;  % sampling frequency
D = 10;    % 10 sec data
N = D*fs;  % number of samples
T = 1/fs;  % sampling period
t = [T:T:D]; % time vector


pure_ecg=ml2;
N=length(pure_ecg);
%=========Figure1==================
figure
subplot(4,1,1); plot(t(1:1800),pure_ecg(1:1800));grid;
title('Pure ECG'); ylabel('Amplitude [mV]');
subplot(4,1,2); plot(t(1:1800),noise_bw(1:1800)); grid;
title('Baseline wander noise'); ylabel('Amplitude
[mV]');
subplot(4,1,3); plot(t(1:1800),noise_ma(1:1800)); grid;
title('Muscle Artifacts'); ylabel('Amplitude [mV]');
subplot(4,1,4); plot(t(1:1800),noise_em(1:1800)); grid;
title('Electrode Motion noise'); ylabel('Amplitude
[mV]');
xlabel('Time [sec]');



%==========Adding Noise to ECG ============
SNR_dB= -10; % SNR for noisy signal
noise=noise_bw + noise_ma + noise_em;
[noisy_ecg,new_SNR] = add_noise(pure_ecg,SNR_dB,noise);
```

```
%=========Figure2===================
figure
subplot(2,1,1); plot(t(1:1800),pure_ecg(1:1800));grid;
title('Pure ECG'); ylabel('Amplitude [mV]');
subplot(2,1,2); plot(t(1:1800),noisy_ecg(1:1800));
grid;
title('Noisy ECG at -10dB'); ylabel('Amplitude [mV]');
xlabel('Time [sec]');


%======= ANC 1 ===========================
[w1 y1 e1]=LMS_algorithm(noise_bw,noisy_ecg,0.001,5);


%======= ANC 2 =====================
[w2 y2 e2]=LMS_algorithm(noise_ma,e1,0.001,5);


%======= ANC 3 =======================
[w3 y3 e3]=LMS_algorithm(noise_em,e2,0.001,5);


%======= Figure 2 ================
figure
subplot(4,1,1); plot(t(1:1800),noisy_ecg(1:1800));grid;
title('Noisy ECG'); ylabel('Amplitude [mV]');
subplot(4,1,2); plot(t(1:1800),e1(1:1800)); grid;
title('ECG after BW noise removed'); ylabel('Amplitude
[mV]');
subplot(4,1,3); plot(t(1:1800),e2(1:1800)); grid;
title('ECG after MA noise removed'); ylabel('Amplitude
[mV]');
subplot(4,1,4); plot(t(1:1800),e3(1:1800)); grid;
title('ECG after EM noise removed'); ylabel('Amplitude
[mV]');
xlabel('Time [sec]');


%========= fidelity Parameters======
[SNRpre1, SNRpost1, MSE1,
ME1]=fidelity_parameters_ecg(pure_ecg, noisy_ecg, e1);
[SNRpre2, SNRpost2, MSE2,
ME2]=fidelity_parameters_ecg(pure_ecg, e1, e2);
[SNRpre3, SNRpost3, MSE3,
ME3]=fidelity_parameters_ecg(pure_ecg, e2, e3);

fprintf('SNRin      SNRout      MSE         ME\n');
disp([SNRpre1, SNRpost1, MSE1, ME1])
disp([SNRpre2, SNRpost2, MSE2, ME2])
disp([SNRpre3, SNRpost3, MSE3, ME3])

result=[SNRpre1, SNRpost1, MSE1, ME1;
        SNRpre2, SNRpost2, MSE2, ME2;
        SNRpre3, SNRpost3, MSE3, ME3];
```

2.3.3 Observation

After the execution of the model over ECG data with LMS algorithm, the following observations are easily captured through plotting of input and output signals. Figure 2.19 shows how the original structure of ECG signal is returned after the use of adaptive filter. For quantitative evolution of adaptive filters with different algorithm, fidelity parameters are calculated and listed in Tables 2.5 and 2.6, for LMS and NLMS algorithms, respectively.

For observation of results, fidelity parameters are calculated. In this case, SNR is calculated using Eq. (2.6) only. With the same equation, SNR is calculated two times, before filtering and after filtering. Before filtering, pure signal and noisy signal are used to calculate SNR as SNR_{pre}. After filtering, pure signal and filtered signal are used to calculate SNR at SNR_{post}. Here, the method of calculation of fidelity parameter gets changed, because pure signal is available. For this, the fidelity parameter function file is changed as provided in fidelity_parameters_ecg.m.

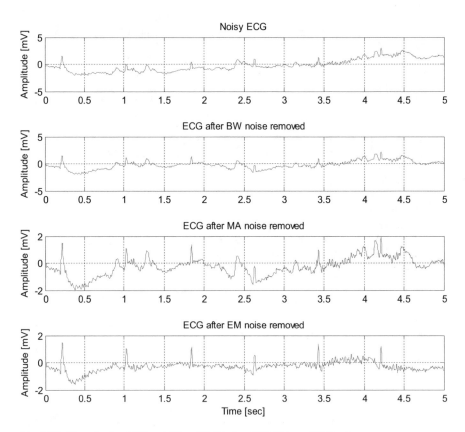

Fig. 2.19 Plot of noisy ECG and filtered ECG after BW, MA, and EM noise

Table 2.5 Performance of ANC based on LMS algorithm for ECG filtering at −10 dB noise

Patient record	ANC	SNR$_{pre}$	SNR$_{post}$	MSE	ME
ml100	ANC1	−10	−7.12822	6.78E−01	2.215343
	ANC2	−7.12822	−6.25934	5.55E−01	2.013185
	ANC3	−6.25934	0.743919	1.11E−01	1.264351
ml101	ANC1	−10	−7.13615	7.09E−01	2.294743
	ANC2	−7.13615	−6.25591	5.79E−01	2.080487
	ANC3	−6.25591	0.876521	1.12E−01	1.2764
ml102	ANC1	−10	−7.18878	3.86E−01	1.655509
	ANC2	−7.18878	−6.34664	3.18E−01	1.498575
	ANC3	−6.34664	0.573811	6.47E−02	1.070724
ml103	ANC1	−10	−7.10142	9.14E−01	2.520933
	ANC2	−7.10142	−6.2603	7.53E−01	2.281012
	ANC3	−6.2603	0.915911	1.44E−01	1.518517
ml104	ANC1	−10	−7.11429	9.80E−01	2.586741
	ANC2	−7.11429	−6.21462	7.97E−01	2.285241
	ANC3	−6.21462	0.461013	1.71E−01	1.616606

Table 2.6 Performance of ANC based on LMS algorithm for ECG filtering at −8 dB noise

Patient record	ANC	SNR$_{pre}$	SNR$_{post}$	MSE	ME
ml100	ANC1	−8	−5.22036	4.37E−01	1.795077
	ANC2	−5.22036	−4.35983	3.58E−01	1.640893
	ANC3	−4.35983	2.390569	7.57E−02	0.98576
ml101	ANC1	−8	−5.23675	4.58E−01	1.865471
	ANC2	−5.23675	−4.36267	3.74E−01	1.700218
	ANC3	−4.36267	2.552293	7.61E−02	0.991838
ml102	ANC1	−8	−5.2747	2.49E−01	1.34043
	ANC2	−5.2747	−4.44271	2.05E−01	1.219463
	ANC3	−4.44271	2.355323	4.29E−02	0.86186
ml103	ANC1	−8	−5.15957	5.84E−01	2.031537
	ANC2	−5.15957	−4.32711	4.82E−01	1.847482
	ANC3	−4.32711	2.884956	9.16E−02	1.194101
ml104	ANC1	−8	−5.18253	6.28E−01	2.080581
	ANC2	−5.18253	−4.27766	5.10E−01	1.836861
	ANC3	−4.27766	2.153018	1.16E−01	1.281077

fidelity_parameters_ecg.m

```
function [SNRpre SNRpost MSE
ME]=fidelity_parameters_ecg(in, no, ou)
% in = input or desired signal
% no = noise signal
% ou = output signal
%---------SNRpre-------------------
inp=0;
for i=1:length(in)
    inp= inp+ in(i)*in(i);
end;
diff1=0;
for i=1:length(no);
    diff1=diff1+((in(i)-no(i))*(in(i)-no(i)));
end;
SNRpre=10*log10(inp/diff1);

%---------SNRpost-------------------
inp=0;
for i=1:length(in)
    inp= inp+ in(i)*in(i);
end;
diff2=0;
for i=1:length(ou);
    diff2=diff2+((in(i)-ou(i))*(in(i)-ou(i)));
end;
SNRpost=10*log10(inp/diff2);

%----------MSE--------------------
MSE=(1/length(ou))*diff2;
%-----------ME--------------------
di=0;
for i=1:length(ou);
    di(i)=abs(((ou(i)-in(i))));
end;
ME=max(di);
```

The results of simulation at noise level at -10 and -8 dB are listed in Tables 2.5 and 2.6, respectively. In these simulations, LMS algorithm is used in ANC.

The observation from Tables 2.5 and 2.6 is very clear to see how SNR gets improved by each ANC used in descendant manner. Patient-wise averaging of SNR gains is plotted in Fig. 2.20. From this, it is observed that ANC performance is better at -10 dB noise level as compared to -8 dB noise level. Performance of NLMS algorithm can also be measured just by replacing the LMS algorithm function by NLMS algorithm. In the same above-discussed simulation, the level of SNR can also be changed easily to test the performance of ANC at different noise levels. Following the same track, implementation of recursive least square (RLS) algorithm can

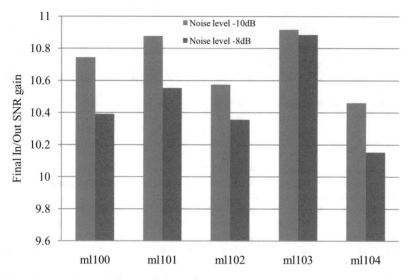

Fig. 2.20 Patient-wise averaging of SNR gains

also be done to further improve the quality of signal. Further using the same concepts, the above simulations can also be used for different biomedical signal and also to other fields of signal processing.

2.4   Adaptive Filter as Optimization Problem

Now, looking from a different angle, the overall focus of adaptive filters is on minimization of error between the desired and output signal. This minimization of error can also be done through other algorithm. Modern optimization methods are also used in place of gradient-based algorithms. For this, the objective function must be defined as error minimization function [5–7]. After this, any modern optimization algorithm can be easily used for the minimization of objective function. The simplest cost function is mean square error function, as given in Eq. (2.9)

$$J_i(n) = \frac{1}{N} \sum_{j=1}^{N} \left[e_{ji}(n) \right]^2 \tag{2.9}$$

where N is the number of the samples of input data and $e_{ji}(n)$ is the jth error for the ith solution. The use of modern optimization technique and their fundamentals are discussed in the next chapter.

References

1. P.S.R. Diniz, *Adaptive Filtering: Algorithms and Practical Implementation* (Springer Science Business Media, Berlin, 2008)
2. A.D. Poularikas, Z.M. Ramadan, *Adaptive Filtering Primer with MATLAB* (CRC, Boca Raton, 2006)
3. B. Widrow, Adaptive filters, in *Aspects of Network and System Theory*, ed. by R. Kalman, N. DeClaris, (Holt, Rinehart, and Winston, New York, 1971), pp. 563–587
4. B. Widrow, J.R. Glover Jr., J.M. McCool, J. Kaunitz, C.S. Williams, R.H. Hearn, J.R. Zeidler, E. Dong Jr., R.C. Goodlin, Adaptive noise cancellation: principles and applications, in *Proceedings of IEEE*, vol. 63 (1975), pp. 639–652
5. M.K. Ahirwal, A. Kumar, G.K. Singh, EEG/ERP adaptive noise canceller design with controlled search space (CSS) approach in cuckoo and other optimization algorithms. IEEE/ACM Trans. Comput. Biol. Bioinfor. **10**(6), 1491–1504 (2013)
6. M.K. Ahirwal, A. Kumar, G.K. Singh, Adaptive filtering of EEG/ERP through Bounded Range Artificial Bee Colony (BR-ABC) algorithm. Digital Signal Process. **25**, 164–172 (2014)
7. M.K. Ahirwal, A. Kumar, G.K. Singh, Descendent adaptive noise cancellers to improve SNR of contaminated EEG with gradient based and evolutionary approach. Int. J. Biomed. Eng. Technol. **13**(1), 49–68 (2013)
8. Y. Ichimaru, G.B. Moody, Development of the polysomnographic database on CD-ROM. Psychiatry Clin. Neurosci. **53**(2), 175–177 (1999)
9. G.B. Moody, R.G. Mark, The impact of the MIT-BIH arrhythmia database. IEEE Eng. Med. Biol. **20**(3), 45–50 (2001). PMID: 11446209
10. G.B. Moody, W.E. Muldrow, R.G. Mark, A noise stress test for arrhythmia detectors. Comput. Cardiol. **11**, 381–384 (1984)
11. M.R. Kose, M.K. Ahirwal, R.R. Janghel, Descendant adaptive filter to remove different noises from ECG signals. Int. J. Biomed. Eng. Technol. **33**(3), 258–273 (2020)

Chapter 3
Swarm and Evolutionary Optimization Algorithm

3.1 Introduction

In mathematical sense, it is possible to define an optimization problem in generic form. Components of optimization problem are design or decision variables, objective function, search space, solution space, and constrains. Different types of optimization problem can be defined based on the combination of optimization components. Decision variables are the solution values within the defined search space and constrain over which objective function is evaluated to get solution space. Effectiveness of the optimization algorithm is based on the selection of decision variables from the search space and to get the best possible solution in solution space within the minimum attempts.

In the beginning, algorithms for optimization were derived from gradient and derivative-based methods. In 1939, Kantorovich was the first to develop an algorithm for linear constraints and the linear objective functions known as linear programming. Later on, different algorithms have been developed for nonlinear optimization. Nonlinearity and multimodality are the main problems in optimization of those objective functions, which do not have their derivates or not defined. Another even more challenging problem arises, when the number of decision variables increases. To cope with these problems, heuristic and meta-heuristic algorithms have been developed, and most of the algorithms are nature-inspired or bio-inspired. These algorithms use the trial-and-error, adaptation, and learning-based approaches to solve the problems, considering them as they cannot find the best solution all the time, but expect them to find the good enough solutions. After 1990, many meta-heuristics/swarm intelligence (SI)/evolutionary technique (ET)-based algorithms have been developed.

At present, optimization of emerging technologies has benefited the different engineering applications and related problems. The increasing complexity of problems motivates the researchers to find out possible ways/methods of getting the

© The Author(s) 2021
M. K. Ahirwal et al., *Computational Intelligence and Biomedical Signal Processing*,
SpringerBriefs in Electrical and Computer Engineering,
https://doi.org/10.1007/978-3-030-67098-6_3

solutions of problems. Utilization of swarm intelligence/evolutionary techniques is an easy way to get the solutions of complex problem. In the next subsection, some basic and fundamental concepts of optimization are discussed.

3.2   Basics of Optimization Problem

Optimization problem in the generic form can be written as given in Eqs. (3.1)–(3.3):
   Minimize:

$$X \epsilon R^n \; f_i(x), \quad (i = 1, 2, \ldots, M), \tag{3.1}$$

   Subject to:

$$\Phi_j(x) = 0, \quad (j = 1, 2, \ldots, J), \tag{3.2}$$

$$\Psi_k(x) \leq 0, \quad (k = 1, 2, \ldots, K), \tag{3.3}$$

where, $f_i(x)$, $\Phi_j(x)$, and $\Psi_k(x)$ are the functions of design vector. $X = (x_1, x_2, \ldots, x_n)$ are called design or decision variables, $f_i(x)$ is called objective function ($i = 1, 2, \ldots, M$). If $M = 1$, there is a single objective. R^n is the search space for decision variables. The space covered by the objective function values is known as solution space; these values are also called feasible solutions. The type of objective function may be linear or nonlinear function. Equalities and inequalities for Φ_j and ψ_k are called constrains [1, 2].

3.2.1   Types of Optimization Problems

1. Linearly constrained problem: Constraints Φ_j and ψ_k are of linear type.
2. Linear programming problem: Both the constraints and objective functions are linear.
3. Nonlinear optimization problem: All f_i, Φ_j and ψ_k are nonlinear.
4. No-constraint problem: A special class of optimization, without any constraints ($j = k = 0$).
5. Feasibility problem: An exceptional case, when there is no objective function, only several constraints are there.

In optimization, main problems are because of nonlinearity and multimodality nature of the objective functions. Another problem is due to high dimensionality of decision variable, and sometimes it is very large ($n = 100$ or 1000).

3.2.2   Classification of Optimization Problems

Optimization problems can be classified into several classes. A brief description is provided here. Figure 3.1 graphically illustrates the classes of optimization problems. Optimization problem may have single ($M = 1$) or multiple ($M > 1$) objectives. Constraints may be equality constraint (if $K = 0$ and $J > 1$) or inequality constraint (when $J = 0$ and $K > 1$). Some problems are unconstrained which do not have any constraints. Optimization functions may be linear or nonlinear, depending upon its equation. In unimodel objective functions, there is a single valley or peak corresponding to single global optimum solution, and hence it is also known as unimodal landscape problem. Unimodal optimization problems have a special class of convex optimization. In case of multiple valley and peaks, objective function is known as multimodal optimization problem. In this type of problem, more than one local maxima and minima are present in the landscape. When the values of all design/decision variables are discrete; in that case, the optimization problem is

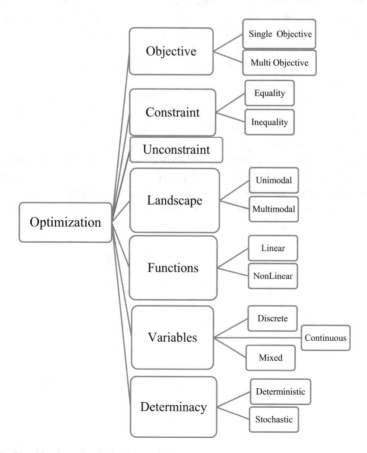

Fig. 3.1 Classification of optimization problems

known as discrete optimization. When the values of decision variables are integers, then it is called as integer programming problem. Graph theory, routing, traveling sales man, vehicle routing, airline scheduling, knap-sack problem, and minimum spanning tree are some common examples of integer programming. In case of continuous optimization problem, all the decision variables take continuous or real values. Problem having both integer and continuous values is known as mixed variable optimization problem. Uncertainty and noise in the design variables, objective functions, and/or constraints convert the problem into stochastic optimization problem, otherwise it remains deterministic optimization problem [1, 2].

3.2.3   Classification of Algorithms

Figure 3.2 shows the classification of optimization algorithm. Deterministic algorithms are those algorithms, which will follow the same path, if the starting point is the same, whether you run the program today or tomorrow. Hill-climbing algorithm is an example of deterministic algorithm. In case of stochastic algorithms, in each run, solutions will be different. But in the final result, there may not be much significant difference. There is always some randomness in these algorithms; and genetic algorithm is a good example.

All the classical and conventional optimization algorithms have deterministic nature. Gradient information is also used in some of the deterministic optimization algorithms, these algorithms are known as gradient-based algorithms. They work fine for smooth and unimodal optimization problems. For complex and difficult multi-objective function, gradient-based algorithms are not suitable. The derivate and gradient information is not required by non-gradient or gradient-free algorithms. Meaning of the word "heuristic" is "to find" or "to discover" by using trial-and-error approach. It is assumed that heuristic algorithms work fine based on their history and

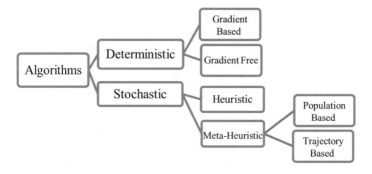

Fig. 3.2 Classification of algorithms

literature, but there is no guarantee for every problem. Development of these heuristic algorithms further opens a new class known as meta-heuristic algorithms. It is also necessary to notice that many definitions are available for heuristics and meta-heuristic algorithms in literature. Most of the meta-heuristic algorithms are inspired from the nature and evolution/survival of species, therefore known as evolutionary techniques. These algorithms are also classified into population-based and trajectory-based algorithms. Genetic algorithms (GA) and particle swarm optimization (PSO) use the concept of chromosomes and particles as multiple agents and are the examples of population-based algorithm. Simulated annealing works on the concept of a single agent/solution that searches for optimal solution. Simulated annealing follows steps or moves to trace the trajectory in the search space.

3.3 Swarm Intelligence/Evolutionary Techniques

Swarm intelligence/evolutionary techniques (ETs) are stochastic and population-based algorithms, which are inspired by the natural process of organic evolution such as mutation, recombination, and selection with specific constraints inside the specific system. ETs follow the idea of survival of the fittest, which means only the best solutions are selected for further process of optimization. On the other hand, swarm intelligence mimics the intelligence of swarms of various species like birds, ants, wolf, and fish. These techniques follow the principle of social interaction and help to reach the best food sources or their destination. Main steps of algorithm are iterated many times, until near or approximate optimal solution is reached. The fundamental idea is to update the population in every iteration toward optimal solution by taking the help of possible solutions [1–3].

3.3.1 Extract and Common Procedure in SI/ET

Now, the common procedure that is followed in each and every swarm-based and evolutionary algorithm is initial population generation, its modification, and selection of the best solution. The modification and selection process of solutions are governed by a specific process that is different for each algorithm. The generalized procedure is presented as a pseudo code given below [4]:

ET/ET

$\text{Initialization}\begin{cases} \textit{Define Objective function} \\ \textit{Initialize population of n solutions} \end{cases}$

  WHILE (Terminiation Condition)

    $\rightarrow t = t + 1$ [*pseudo time or iteration counter*]

    FOR (*loop over all solutions and all D dimensions*)

        $\Rightarrow$ Generate new solutions/modification in population

        $\Rightarrow$ *fitness calculation*

        $\Rightarrow$ replace old solutions, if new solutions found better/according to selection criteria

    END FOR

    $\rightarrow$ *Find fittest among all solution*

  END WHILE

  Output the final results}

After getting the overview and common procedure followed by SI/ET, the difference between various SI/ET can be easily understood just by observing their population modification/updation method and selection criteria. Initial population generation is same in all the algorithms. Population of solution is called by different name in each of the algorithm, like chromosome, food source, and particle position. The very confined description of various SI/ET is given below.

3.3.2   Genetic Algorithm

This is one of the famous and oldest ETs developed in 1975. Genetic algorithm uses two main operations: crossover and mutation operation. Here, population of solutions is called as chromosomes. Solutions are modified by crossover and mutation operation [5]. Cost/objective function is used to calculate the fitness value to each chromosome. Generally, selection of new chromosomes are done by Roulette Wheel method. In case of real valued GA, the chromosomes are real number, while in case of binary GA (the famous and conventional one), they use binary numbers as chromosomes. Here, the operations used in real GA are described with their Eq. (3.6). Equation (3.4) is used for crossover operation.

$$C_i^{\text{gen}+1} = a \cdot (C_i^{\text{gen}}) + (1-a) \cdot C_j^{\text{gen}} \quad \text{and} \quad C_j^{\text{gen}+1}$$
$$= (1-a) \cdot (C_i^{\text{gen}}) + a \cdot C_j^{\text{gen}}, \tag{3.4}$$

In the above equation, C_i^{gen} and C_j^{gen} represent two random chromosomes. $C_i^{\text{gen}+1}$ and $C_i^{\text{gen}+1}$ represent the next-generation chromosomes. This equation is a linear combination of parent chromosomes. a denotes the random number between 0 and 1. After that, the mutation operation is just the random alteration of chromosomes by multiplying the random real value.

3.3.3 Artificial Bee Colony Algorithm

This is a famous SI-based algorithm and has been proposed in 2005. Various applications of this SI optimization algorithm were found in different engineering and science optimization problems. This algorithm follows the principle of foraging behavior of a bee in their colony. This algorithm is based on the simulation of behavior of bees in the real world. Three categories of artificial bees are assumed, called as employed bees, onlooker bees, and scout bees. The employed bees have the same population as onlooker bees. The work of the employed bees is to search the food sources, in a predefined closed search space also called as bounded search space. These bees will share the food source information with onlooker bees. The onlooker bees further explore these food sources. When some food sources get abandoned, then the bees associated with that food source are translated to scout bees, and new random food sources are assigned to them [7, 8]. Equation (3.5) is used for population (termed as colony) initialization.

$$X_{i,j} = X_{\min,j} + \text{rand}[0, 1] * \left(X_{\max,j} - X_{\min,j}\right). \tag{3.5}$$

In the above equation, $i = 1, 2, \ldots,$ SN, $j = 1, 2, \ldots, D$. SN and D represent the colony size and its dimension, respectively. $X_{\min,j}$ and $X_{\max,j}$ represent the lower and upper bounds, respectively. Now, in bee initialization state, each employed bee (X_i) searches/generates a new source of food V_i near to its present position through Eq. (3.6).

$$V_{i,j} = X_{i,j} + \phi_{i,j} * \left(X_{i,j} - X_{k,j}\right). \tag{3.6}$$

In the above equation, $k \in [1, 2, \ldots, \text{SN}]$ and $j \in [1, 2, \ldots, D]$ are randomly selected indexes and k must be different from i. $\phi_{i,j}$ is a random number between -1 and 1. Once V_i is obtained, its evaluation and comparison is to be done with X_i on the basis of objective/fitness function. If it is found that V_i is equal or better than that of X_i, then V_i replaces X_i, otherwise X_i remains unchanged. It is called as greedy selection approach for the selection of new solutions in next population.

Now, onlooker bees follow probabilistic selection process to select the food sources. The formula used in this probabilistic selection is given as Eq. (3.7).

$$P_i = \frac{f_i}{\sum_{i=1}^{\text{SN}} f_i}. \tag{3.7}$$

In the above equation, f_i is the fitness value of solution i. Higher the f_i is, more is the probability that ith food source is selected. This P_i value basically evaluates the nectar information gathered by employed bees to select a food source X_i for onlooker bees. After this, modification is done by Eq. (3.6). The food sources which cannot be enhanced in several predefined trials (termed as limit) are called as abandoned, and the bees are known as scout bees. For scout bees, new food sources are assigned randomly.

3.3.4 Particle Swarm Optimization Algorithm

This is another famous SI-based algorithm for solving the optimization problems and was developed in 1995. This algorithm simulates the behavior of swarm of flying bird. This algorithm follows the principle of self-experiences and social experiences. In this, population is represented by flying particles in a search space which tries to move toward a global optimum solution. The overall best solution, called as global best and local best solutions, is identified in each iteration and used for updating the present population, for the next iteration. For each particle, the position and velocity are maintained. Before updating the position of the particle, velocity is updated. Each particle is searching in the search space to find the best solution, for its velocity and position updation. To update the velocity and position of the particle, its own experience as well as experience of best particle is used [9, 10]. This updation process is given in Eq. (3.8).

$$V_i(t+1) = V_i(t) + U(0, \emptyset 1) * (P_i(t) - X_i(t)) + U(0, \emptyset 2) \\ * \left(P_g(t) - X_i(t)\right), \tag{3.8}$$

$$X_i(t+1) = X_i(t) + V_i(t+1). \tag{3.9}$$

In the above equation, t is a time instant and represents iterations. The updated position is $X_i(t+1)$ of the ith particle for the next iteration. Updated velocity vector is $V_i(t+1)$. For PSO, generally S is used to represent the population size of particles, and every particle has D dimensions. $U(0, \varphi i)$ represents random numbers uniformly distributed between 0 and φi, generated randomly in each iteration for every particle. The parameters $\varphi 1$ and $\varphi 2$ represent the magnitude forces in the direction of personal best position (particle best—P_{best}) P_i and neighborhood best position (global best—G_{best}) P_g. The final selection of the updated particle value is done by checking its fitness value.

Due to increasing popularity of PSO applications, different variants of PSO have been developed. Inertia-weighed PSO is one of the famous variants among them. In this, change is done in velocity-updated equation as explained through Eq. (3.10).

$$V_i(t+1) = I * V_i(t) + U(0, \emptyset 1) * (P_i(t) - X_i(t)) + U(0, \emptyset 2) \\ * \left(P_g(t) - X_i(t)\right). \tag{3.10}$$

In the above equation, I represents inertia, which is also updated by the decay function with respect to time as given in Eq. (3.11),

$$I = I_{max} - (I_{max} - I_{min}) * n/N. \tag{3.11}$$

In the above equation, I_{max} and I_{min} represent the upper limit and lower limit of inertia, respectively. N represents the total number of iteration. Current iteration is denoted by n.

3.4 Objective Functions

General description of objective function includes d as number of variables or dimension of problem. Search domain is defined as $R_{min} \leq x_i \leq R_{max}$, where, $i = 1, 2, \ldots, d$, R_{min} and R_{max} are lower and upper limits of variables, respectively. The, number of local minima or local maxima are more than one in multi-model problems. The position of global minima or maxima is denoted as $X*$ and the value of global minima/maxima is $f(X*)$. Some examples of benchmark objective functions are described below, and their MatLab code is also demonstrated with their visualization [1, 2, 11, 12].

Example 1: Sum Square

$$f(x) = \sum_{i=1}^{d} x_i^2. \tag{3.12}$$

Dimension	Search domain	Number of local minima	Position of global minima	Global minima
1	−5 to +5	1	0	0

Example 2: Shifted sum square

$$f(x) = \sum_{i=1}^{n} x_i^2 - 20. \tag{3.13}$$

Dimension	Search domain	Number of local minima	Position of global minima	Global minima
1	−5 to +5	1	0	−20

Example 3: Combination of sine and cosine

$$f(x) = x + 3\sin(2x) + 6\sin(2x) + \cos(2x). \tag{3.14}$$

Dimension	Search domain	Number of local minima	Position of global minima	Global minima
1	−5 to +5	2	−4.3	−11.4213

MatLab code for the above-defined objective functions is given below, and their plots are shown in Figures 3.3, 3.4, and 3.5.

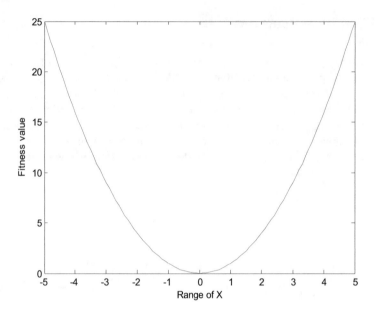

Fig. 3.3 Plot of sum square function

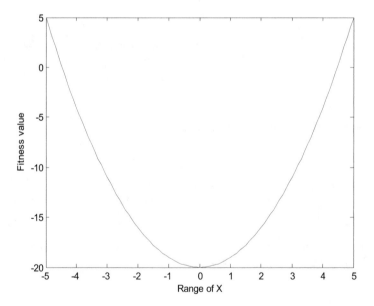

Fig. 3.4 Plot of shifted sum square function

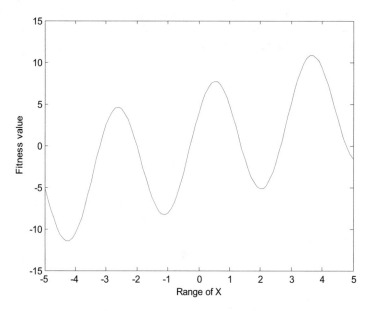

Fig. 3.5 Plot of combination of sin and cos functions

One_D_Benchmarks.m

```
clc;
clear all;

% Simple example of Objective functions
% ===============================================

% Objective function 1 (Sphere Function)
x=-5:0.1:5; % Range
res=x.^2;
figure
plot(x,res) % plot of function
xlabel('Range of X')
ylabel('Fitness value')




% Objective function 2 (Shifted Sphere Function)
x=-5:0.1:5; % Range
b=-20;
res=x.^2 + b;
figure
plot(x,res) % plot of function
xlabel('Range of X')
ylabel('Fitness value')
```

```
% Objective function 3 (combination of sin and cos
function)
x=-5:0.1:5; % Range
res=x+3*cos(2*x)+6*sin(2*x)+cos(2*x);
figure
plot(x,res) % plot of function
xlabel('Range of X')
ylabel('Fitness value')
```

Now, some examples of two-dimensional objective functions are also discussed below:

Example 4: Sphere

$$f(x) = \sum_{i=1}^{n} x_i^2.$$ (3.15)

Dimension	Search domain	Number of local minima	Position of global minima	Global minima
2	−5 to +5	1	0,0	0

Example 5: Ackley

$$f(x_1, x_2) = -20e^{\left[-0.2\sqrt{0.5\left(x_1^2 + x_2^2\right)}\right]} - e^{[0.5(\cos 2\pi x_1 + \cos 2\pi y_1)]} + e + 20.$$ (3.16)

Dimension	Search domain	Number of local minima	Position of global minima	Global minima
2	−5 to +5	Several	0,0	0

Example 6: Michalewics

$$f(x) = -\sum_{i=1}^{d} \sin(x_i) \sin^{2m}\left(\frac{ix_i^2}{\pi}\right).$$ (3.17)

Dimension	Search domain	Number of local minima	Position of global minima	Global minima
2	0, π	Several	2.20, 1.57	−1.8013
2	−5 to +5	Several	−5, 1.6	−1.8263

In the above function, value of extra parameter *m* is taken as 10.

Example 7: Schwefel $f(x) = 418.9829\,d - \sum_{i=1}^{d} x_i \sin\left(\sqrt{|x_i|}\right).$ (3.18)

Dimension	Search domain	Number of local minima	Position of global minima	Global minima
2	−500 to 500	Several	421, 421	2.7197e−004 (approx 0)
2	−50 to 50	Several	50, 50	767.0797

Example 8: Griewank

$$f(x) = \sum_{i=1}^{d} \frac{x_i^2}{4000} - \prod_{i=1}^{d} \cos\left(\frac{x_i}{\sqrt{i}}\right) + 1.$$ (3.19)

Global minima	Dimension	Search domain	Number of local minima	Position of global minima
2	−600 to 600	Several	0,0	0
2	−50 to −50	Several	0,0	0

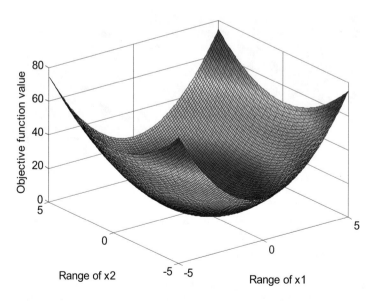

Fig. 3.6 Surface plot of sphere function

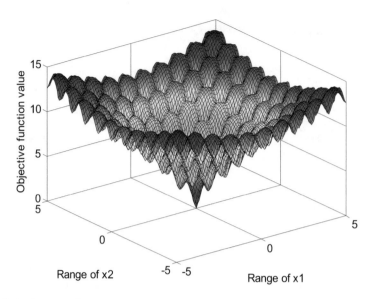

Fig. 3.7 Surface plot of Ackley function

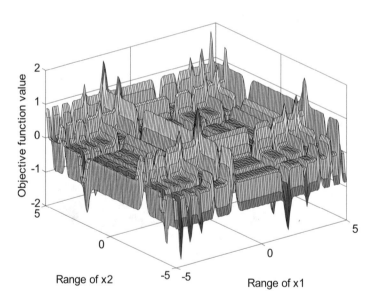

Fig. 3.8 Surface plot of Michalewics function

MatLab codes for the above-defined two-dimensional objective functions are given below, and their surface plots are shown in Figures 3.6, 3.7, 3.8, 3.9, and 3.10.

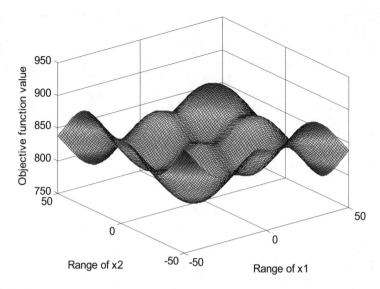

Fig. 3.9   Surface plot of Schwefel function

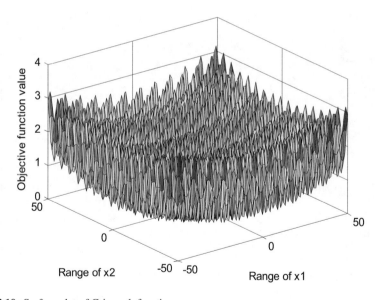

Fig. 3.10   Surface plot of Griewank function

Two_D_Benchmarks.m

```
clc;
clear all;
%------------------------------------------------


% Objective function 1  (Sphare Functions)
x1=-5:0.1:5; % Range
x2=-5:0.1:5; % Range
z=[];

for i=1:length(x1)
    for j=1:length(x2)

    z(i,j) = sphare([x1(i) x2(j)]);

    end
end
figure
surf(x1,x2,z) % plot of function
xlabel('Range of x1')
ylabel('Range of x2')
zlabel('Objective function value')


%------------------------------------------------


% Objective function 2 (Ackley Function)
x1=-5:0.1:5; % Range
x2=-5:0.1:5; % Range
z=[];

for i=1:length(x1)
    for j=1:length(x2)

    z(i,j) = ackley([x1(i) x2(j)]);

    end
end
figure
surf(x1,x2,z) % plot of function
xlabel('Range of x1')
ylabel('Range of x2')
zlabel('Objective function value')


%------------------------------------------------


% Objective function 3 (Michalewics Function)
x1=0:0.1:pi; % Range
x2=0:0.1:pi; % Range
z=[];
```

```
for i=1:length(x1)
    for j=1:length(x2)

    z(i,j) = mich([x1(i) x2(j)]);

    end
end
figure
surf(x1,x2,z) % plot of function
xlabel('Range of x1')
ylabel('Range of x2')
zlabel('Objective function value')

%---------------------------------------------

% Objective function 4 (Schwefel Functions)
x1=-50:1:50; % Range
x2=-50:1:50; % Range
z=[];

for i=1:length(x1)
    for j=1:length(x2)

    z(i,j) = schw([x1(i) x2(j)]);

    end
end
figure
surf(x1,x2,z) % plot of function
xlabel('Range of x1')
ylabel('Range of x2')
zlabel('Objective function value')

%---------------------------------------------

% Objective function 5 (griewank Functions)
x1=-50:1:50; % Range
x2=-50:1:50; % Range
z=[];

for i=1:length(x1)
    for j=1:length(x2)

    z(i,j) = griewank([x1(i) x2(j)]);

    end
end
figure
surf(x1,x2,z) % plot of function
xlabel('Range of x1')
ylabel('Range of x2')
zlabel('Objective function value')
```

In all the above examples, dimension and range of variables can be easily changed. Commands to find out global minimum with its position in the above examples are given below.

```
>>global_minimum = min(min(z));
>> [p1,p2]=find(z== global_minimum)
>>x1(p1)
>>x2(p2)
```

Above code is also used to find the global maxima, just replacing the minimum function with maximum. These benchmark functions and their variants are listed and described in [11, 12].

3.5 Implementation of Real Genetic Algorithm

For implementation, suppose that the population of chromosomes is $C_1, C_2, \ldots, C_L$, where L is the size of population. Genetic operations are (1) crossover and (2) mutation. After performing crossover operation, population size becomes $2L$. Because, during crossover in every iteration, two chromosomes are randomly selected to generate two new child chromosomes, and this process is repeated for L time. Over these new $2L$ child chromosomes, mutation is performed. If mutation probability is not defined, then all chromosomes go through mutation process. Among these $2L$ mutated chromosomes, selection of L chromosomes is done by Roulette Wheel selection method (other methods are also available for the same). New population is evaluated through objective/fitness function, and normalization is performed over fitness values. Cumulative distribution of the obtained normalized fitness values is also calculated, and finally random number r simulates the rotation of Roulette Wheel. This wheel rotation process is repeated L times to select L chromosomes for the next iteration. Below is the simplified algorithm for the same [6, 13].

Simplified Real GA Algorithm:

Step 1: Generate random initial population with L chromosomes within the range (R_{min} to R_{max}).

Step 2: Select two chromosomes randomly as parents for arithmetic crossover to get two new child chromosomes. This step is repeated L times.

Step 3: Mutation operation is performed over all the $2L$ child chromosomes by random number multiplication.

Step 4: Evaluate all $2L$ mutated chromosomes through fitness function and calculate cumulative distribution of normalize fitness values.

Step 5: Generate random number to simulate Roulette Wheel selection to select L fittest chromosomes for next iteration.

Step 6: Identify the best fitness value with its chromosome.

Step 7: Repeat process of steps 2–6 as per stopping criterion (maximum iteration).

A simplified code, based upon the above algorithm, is given below with comments to explain the entire code.

Real_GA.m

```
% Real Genetic Algorithm
clc;
clear all;
close all;

% Step 1: Generate the initial population of L chro-
mosomes within range.
% where L=10 and Range is 0 to 9

Rmin=-5;  % lower limit
Rmax=5;   % Upper limit
L=10;
pop=Rmin:0.001:Rmax;            % population with
resolution of 0.001 (total 9001 values)
pos=round(rand(1,L)*9000)+1;   % L random Indexes
between 1 to 9000
chromosome=pop(pos);           % value at random po-
sitions (obtaining values from indexes)

iteration = 50;
tic;
for i=1:iteration % iteration of GA
chromosome1=[]; % empty matrix for modified chromo-
somes
chromosome2=[]; % empty matrix for selected chromo-
somes after modification
for do=1:1:L % Genetic Operations over whole popula-
tion(this loop runs L times)

    % Step 2: select two chromosomes randomly as
parents for crossover.
    % Perform arithmetic crossover, to generate two
child chromosomes.

    % 2.1 Generate random indexes rx1 and rx2 for
two chromosomes
    rx1=round(rand(1)*9)+1;
    rx2=round(rand(1)*9)+1;

    % 2.2 Random number r1
    r1=rand;

    % 2.3 CrossOver Operation to get child chromo-
somes y1 and y2
    y1=r1*chromosome(rx1)+(1-r1)*chromosome(rx2);
    y2=(1-r1)*chromosome(rx1)+ r1*chromosome(rx2);
```

```
    % Step 3: Mutate all child chromosomes by random
change.
    % 3.1 Random number r1
    r2=rand;

    % 3.2 Multiply child chromosomes with random
number r2 to get z1 and z2
    z1=y1*r2;
    z2=y2*r2;

    % chromosome1=[chromosome1 z1 z2];
    % check for Range
    if ( z1 >= Rmin & z1 <= Rmax )
    chromosome1=[chromosome1 z1 ];
    else
    chromosome1=[chromosome1 chromosome(rx1) ];
    end

    if ( z2 >= Rmin & z2 <= Rmax )
    chromosome1=[chromosome1 z2 ];
    else
    chromosome1=[chromosome1 chromosome(rx2) ];
    end

end

% Step 4: Evaluate entire 2*L population
fitness=objective1(chromosome1);

% 4.1 Obtain cumulative distribution of normalize
fitness values.
nor_fitness=fitness/max(fitness);
cum_sum=cumsum(nor_fitness);
nor_cum_sum=cum_sum/max(cum_sum);
%bar(nor_cum_sum);

%Step 5: Apply Roulette Wheel selection to select
fittest L chromosomes as next generation.
for ii=1:L
% 5.1 Generate random number r3
r3=rand;
% 5.2 Find the Index where r3 falls
Index=find(r3<nor_cum_sum);
% 5.3 Select that chromosome from population
chromosome2=[chromosome2 chromosome1(Index(1))];
```

```
end

% Step 6: find best among chromosome population (fi-
nal weight vector)
fitness_new=objective1(chromosome2);
[V I]=max(fitness_new);
fprintf('\n In iteration =%d the best fitness value
is =%f for chromosome =%f',i,V,chromosome2(I));
convergence(i)=V;

% 6.1 Replacing old chromosomes with new chromosomes
chromosome=chromosome2;



% Visualization of Objective function with Chromo-
somes
curve=objective1(pop);
fit=objective1(chromosome);
plot(pop,curve)
hold on
plot(chromosome,fit,'r*','MarkerSize',15)
getframe;
xlabel('Chromosomes values')
ylabel('Fitness Values')
pause(0.2)
hold off

% Step 7: Repeat Steps 2 to 6 until stopping crite-
rion is not satisfied (maximum iteration).
end
Time=toc;
fprintf('\n Time of execution=%f',Time);
figure
plot(convergence);
xlabel('Iterations')
ylabel('Fitness Values')
title('Convergence of GA for objective1');
```

While using this code for evaluation of algorithm comment or remove the pause command.

The objective functions 1 and 2 are taken for maximization in the above code. Their code, details, and plots (Figures 3.11 and 3.12, respectively) are given below.

$$f(x) = x + 10 \sin(2x) - 3 \cos(4x) + \sin(x). \tag{3.20}$$

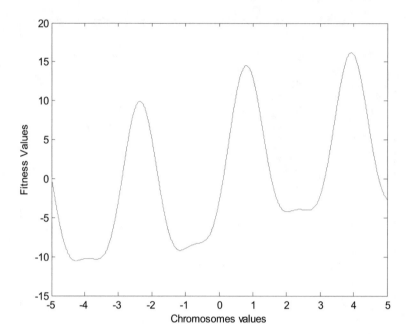

Fig. 3.11 Plot of objective 1 function

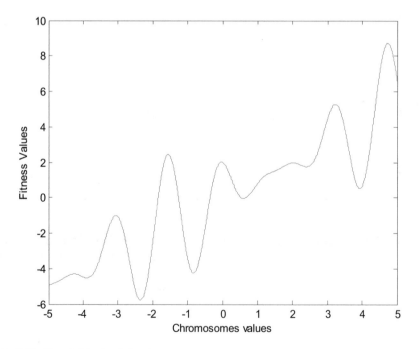

Fig. 3.12 Plot of objective 2 function

objective1.m
```function [res]=objective1(x)```
```res=x+10*sin(2*x)-3*cos(4*x)+sin(x);```

Detail of objective 1 function

Dimension	Search domain	Number of local maxima	Position of global maxima	Global maxima
1	−5 to +5	2	3.9300	16.2204

Another objective 2 function is,

$$f(x) = x + \sin(3x) - 3\cos(4x) - \sin(5x). \tag{3.21}$$

Objective2.m
```function [res]=objective2(x)```
```res=x+sin(3*x)+2*cos(4*x)-sin(5*x);```

Detail of objective 2 function

Dimension	Search Domain	Number of local maxima	Position of global maxima	Global maxima
1	−5 to +5	2	4.7280	8.7200

3.5.1 Visualization of Chromosome over Plot of Objective Function

For objective 1 function, in the above plot (Fig. 3.13), red stars show the fitness values of chromosomes with positions corresponding to x axis, some of them are overlapping.

For objective 2 function, same illustration is used (Fig. 3.14). After several iterations, these chromosomes will move toward the global optima, and on running the code, it can be easily visualized.

3.5.2 Performance Measure of Optimization Algorithm

Convergence plot: The convergence plot shows the progress of algorithm for each iteration toward the optimal solution, called as converging to the optimal solution. Figures 3.15 and 3.16 show the convergence plots of the above-described GA for objective 1 and objective 2 functions.

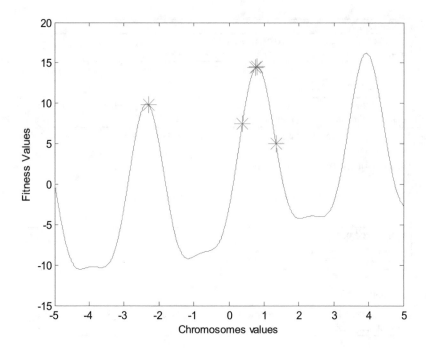

Fig. 3.13 Plot of objective 1 with chromosomes

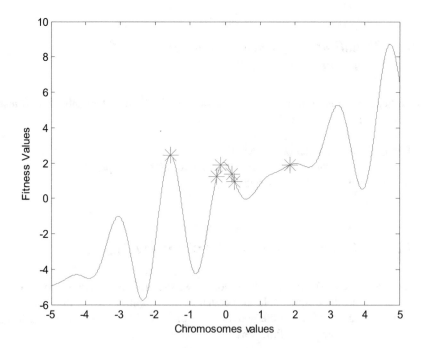

Fig. 3.14 Plot of objective 2 with chromosomes

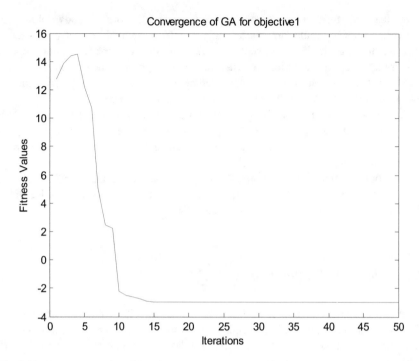

Fig. 3.15 Convergence plot of GA for objective 1

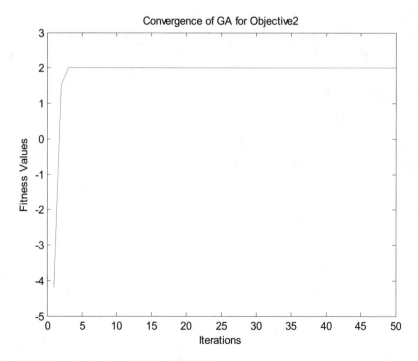

Fig. 3.16 Convergence plot of GA for objective 2

In each trial, different convergence plot is obtained. From the above plots, it is observed that GA could not successfully reach to optimal values of objective functions used. From the above plots, it is also observed that the GA is not properly converging to optimal solutions as listed in the description of the objective function details. In such case, some modification can be done to enhance the performance of GA. Demonstration of modification is provided in a later section.

Time of execution: Analysis of execution time of algorithm is also required; this is dependent on the equations used in population updating and selection of fittest solution. Following notations and calculations can be used for time analysis purpose as listed in Table 3.1 [4].

For both the objective functions, T_0, T_1, and T_2 are calculated for 50 iterations with ten population size for four trials. Table 3.2 is listed with execution time. Further, in Table 3.3, T_0, T_1, and T_2 are calculated.

3.5.3   Simple Modification in Real GA

For modification in GA or in any other optimization algorithm, there are many ways, like modification in population updating equations or use of new operators, use of new selection scheme to select the updated solution for next iteration and method for random initialization of population.

Table 3.1 Execution time notations

Notation of running time	Running time of algorithms
$T_0 =$ Code execution time	$T_0 = \frac{\text{Total time}}{\text{No. of iteratinos} \times \text{Population size}}$
$T_1 =$ Time of n evaluations	$T_1 = \frac{\text{Total time}}{\text{No. of iteration}}$
$T_2 =$ Mean of N trials	$T_2 =$ Mean of N trials

Table 3.2 Execution time for different trials

Trial	Total time for objective 1	Total time for objective 2
1	1.174565	1.182849
2	1.219689	1.233748
3	1.198759	1.199289
4	1.223773	1.255916

Table 3.3 Execution time (in seconds) analysis of GA

Time notations	Objective 1	Objective 2
T_2	1.204197	1.217951
T_1	0.024084	0.024359
T_0	0.002408	0.002436

To make it simpler to understand, here simple modification of mutation operation is demonstrated. For that, a concept of mutation probability is added and its value is set to 0.2 or 0.3. In code of GA, only few lines are modified under the section of mutation operation, and the rest of the code is the same.

```
Modified_Real_GA.m
% Modified Real Genetic Algorithm
clc;
clear all;
close all;

% Step 1: Generate the initial population of L
chromosomes within range.
% where L=10 and Range is 0 to 9

Rmin=-5;   % lower limit
Rmax=5;   % Upper limit
L=10;
pop=Rmin:0.001:Rmax;              % population with
resolution of 0.001 (total 9001 values)
pos=round(rand(1,L)*9000)+1;   % L random Indexes
between 1 to 9000
chromosome=pop(pos);             % value at random
positions (obtaining values from indexes)
P=0.2; %Mutation Probability
iteration = 50;
tic;
for i=1:iteration % iteration of GA
chromosome1=[]; % empty matrix for modified chromo-
somes
chromosome2=[]; % empty matrix for selected chromo-
somes after modification
for do=1:1:L % Genetic Operations over whole popu-
lation(this loop runs L times)

    % Step 2: select two chromosomes randomly as
parents for crossover.
    % Perform arithmetic crossover, to generate two
child chromosomes.

    % 2.1 Generate random indexes rx1 and rx2 for
two chromosomes
    rx1=round(rand(1)*9)+1;
    rx2=round(rand(1)*9)+1;

    % 2.2 Random number r1
    r1=rand;

    % 2.3 CrossOver Operation to get child chromo-
somes y1 and y2
```

```
    y1=r1*chromosome(rx1)+(1-r1)*chromosome(rx2);
    y2=(1-r1)*chromosome(rx1)+ r1*chromosome(rx2);

    % Step 3: Mutate all child chromosomes by ran-
dom change.
    % 3.1 Random number r1
    r2=rand;

    % 3.2 Multiply child chromosomes with random
number r2 to get z1 and z2
    % Modification (if mutation probability is <=
P).
    if (r2<=P)
        z1=y1*r2;
        z2=y2*r2;
    else
        z1=y1;
        z2=y2;
    end


    % chromosome1=[chromosome1 z1 z2];
    % check for Range
    if ( z1 >= Rmin & z1 <= Rmax )
    chromosome1=[chromosome1 z1 ];
    else
    chromosome1=[chromosome1 chromosome(rx1) ];
    end

    if ( z2 >= Rmin & z2 <= Rmax )
    chromosome1=[chromosome1 z2 ];
    else
    chromosome1=[chromosome1 chromosome(rx2) ];
    end

end

% Step 4: Evaluate entire 2*L population
fitness=objective1(chromosome1);

% 4.1 Obtain cumulative distribution of normalize
fitness values.
nor_fitness=fitness/max(fitness);
cum_sum=cumsum(nor_fitness);
nor_cum_sum=cum_sum/max(cum_sum);
```

```
%bar(nor_cum_sum);

%Step 5: Apply Roulette Wheel selection to select
fittest L chromosomes as next generation.
for ii=1:L
% 5.1 Generate random number r3
r3=rand;
% 5.2 Find the Index where r3 falls
Index=find(r3<nor_cum_sum);
% 5.3 Select that chromosome from population
chromosome2=[chromosome2 chromosome1(Index(1))];
end

% Step 6: find best among chromosome population
(final weight vector)
fitness_new=objective1(chromosome2);
[V I]=max(fitness_new);
fprintf('\n In iteration =%d the best fitness value
is =%f for chromosome =%f',i,V,chromosome2(I));
convergence(i)=V;

% 6.1 Replacing old chromosomes with new chromo-
somes
chromosome=chromosome2;


% Visualization of Objective function with Chromo-
somes
curve=objective1(pop);
fit=objective1(chromosome);
plot(pop,curve)
hold on
plot(chromosome,fit,'r*','MarkerSize',15)
getframe;
xlabel('Chromosomes values')
ylabel('Fitness Values')
%pause(0.2)
hold off

% Step 7: Repeat Steps 2 to 6 until stopping crite-
rion is not satisfied (maximum iteration).
end
Time=toc;
fprintf('\n Time of execution=%f',Time);
figure
plot(convergence);
xlabel('Iterations')
ylabel('Fitness Values')
title('Convergence of GA for objective1');
```

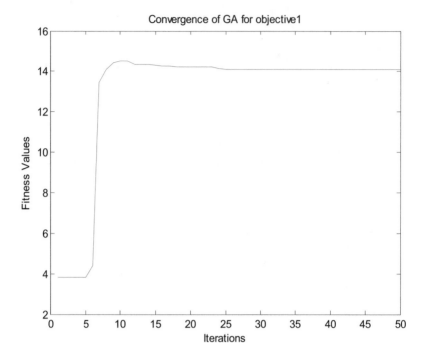

Fig. 3.17 Convergence plot of modified GA for objective 1

After the above modification, improvement in performance is observed and as illustrated in convergence plot for objective 1 as shown in Fig. 3.17. Now, the better value of optimal solution, as compared to previous plot (Fig. 3.15), is achieved.

3.6 Application to Adaptive Filters

The above-demonstrated code is very simple and easy to understand the insight of genetic algorithm. This code can be modified as in any of the above-discussed algorithm, modification is only required at the places of population updation and selection for the next iteration.

After getting all the necessary information about the use of swarm-based optimization techniques, these algorithms can be easily used for the optimization of adaptive filters as discussed in the previous chapter. The concept is very simple; treat mean square error (MSE) of adaptive filter as objective function for minimization. With EEG data and MSE objective function, all these algorithms can be used for optimization of adaptive filters performance [14–16].

References

1. X.-S. Yang, *Nature-inspired metaheuristic algorithms* (Luniver Press, Frome, 2010)
2. X.-S. Yang, *Engineering optimization: an introduction with metaheuristic applications* (Wiley, New York, 2010)
3. A. Chakraborty, A.K. Kar, Swarm intelligence: A review of algorithms, in *Nature-Inspired Computing and Optimization. Modeling and Optimization in Science and Technologies*, ed. by S. Patnaik, X. S. Yang, K. Nakamatsu, vol. 10, (Springer, Cham, 2017). https://doi.org/10.1007/978-3-319-50920-4_19
4. M.K. Ahirwal, A. Kumar, G.K. Singh, EEG/ERP adaptive noise canceller design with Controlled Search Space (CSS) approach in Cuckoo and other optimization algorithms. IEEE/ACM Trans. Comput. Biol. Bioinform. **10**(6), 1491–1504 (2013). https://doi.org/10.1109/TCBB.2013.119
5. M. Srinivas, L.M. Patnaik, Genetic algorithms: A survey. Computer **27**(6), 17–26 (1994)
6. T. Yalcinoz, H. Altun, M. Uzam, Economic dispatch solution using a genetic algorithm based on arithmetic crossover, in *Proceedings of Conference on Porto Power Technology, IEEE*, vol. 2, 2011, pp. 1–4
7. S.A.M. Fahad, M.E. El-Hawary, Overview of Artificial Bee Colony (ABC) algorithm and its applications, in *Proceedings of International Conference on Systems, IEEE*, 2012, pp. 1–6
8. B. Akay, D. Karaboga, A modified artificial bee colony algorithm for real-parameter optimization. Inform. Sci. **192**, 120–142 (2012)
9. J. Kennedy, R.C. Eberhart, Particle swarm optimization, in *Proceedings of international conference on Neural Networks, IEEE*, vol. 4, 1995, pp. 1942–1948
10. R. Poli, J. Kennedy, T. Blackwell, Particle swarm optimization: An overview. Swarm Intell. **1**, 33–57 (2007)
11. S. Surjanovic, D. Bingham, Virtual library of simulation experiments: Test functions and datasets (2013), http://www.sfu.ca/~ssurjano. Accessed 10 Oct 2020
12. M. Jamil, X.-S. Yang, A literature survey of benchmark functions for global optimisation problems. Int. J. Math. Model. Numer. Optim. **4**(2), 150–194 (2013)
13. M.K. Ahirwal, A. Kumar, G.K. Singh, Adaptive filtering of EEG/ERP through noise cancellers using an improved PSO algorithm. Swarm Evol. Comput. **14**, 76–91 (2014)
14. M.K. Ahirwal, A. Kumar, G.K. Singh, Improved range selection method for evolutionary algorithm based adaptive filtering of EEG/ERP signals. Neurocomputing **144**, 282–294 (2014)
15. M.K. Ahirwal, A. Kumar, G.K. Singh, Adaptive filtering of EEG/ERP through bounded range artificial bee colony (BR-ABC) algorithm. Digit. Signal Process. **25**, 164–172 (2014)
16. M.K. Ahirwal, A. Kumar, G.K. Singh, Analysis and testing of PSO variants through application in EEG/ERP adaptive filtering approach. Biomed. Eng. Lett. **2**(3), 186–197 (2012)

Chapter 4
Prediction and Classification

4.1 Introduction

In this chapter, the basic fundamentals of predictions and classification have been discussed with few examples. First of all, the theory of prediction and classification is explained, and then how these techniques are applied to real-world problem is illustrated.

In real world, there are so many applications and problems, where prediction and classification can be applied to get a possible solution. The basic elements for doing prediction or classification are input and output variables. With the help of the number of instances or samples of input and output variables, the model for prediction and classifications is developed/trained or learned. This trained model is used to predict or classify the new or future instances or input variables. This is also called supervised learning approach.

The fundamental principle of prediction is the assumption that the relation between the input and output variable is governed by some hidden model or system. This model can be built/developed by regression analysis. The aim of regression analysis is to find out the coefficients of model for prediction. The calculation of coefficients is done from the available data. In case of classification, these predictions are further projected or classified into classes by activation function/or by some thresholds values. In later sections, at first, prediction and regression are discussed through suitable examples, and then classification process has been explained with real biomedical signals. At the end of this chapter, the development of artificial neural network is also described in a very simple manner.

© The Author(s) 2021
M. K. Ahirwal et al., *Computational Intelligence and Biomedical Signal Processing*,
SpringerBriefs in Electrical and Computer Engineering,
https://doi.org/10.1007/978-3-030-67098-6_4

4.2 Prediction and Regression

4.2.1 Simple Linear Regression

The very first step is to learn the concept of regression by simple linear regression. Many types of regression methods with different algorithm are available like nonlinear, multiple linear, and multiple nonlinear regression. Multiple regression means multiple input variable and single output variable. Simple regression is just about to find the relation between two variables. A very simple example dataset is given below, and over the same dataset, simple linear regression is applied.

In Table 4.1, two variables X and Y are given, they are also called as attributes. The first variable (X) is input or independent variable, and the second variable (Y) is output or dependent variable. Figure 4.1 provides the visualization of the dataset.

The mathematic equation for simple linear regression is given in Eq. (4.1).

Table 4.1 Sample of an example dataset

X (input)	Y (output)
2	2
4	3
6	5
8	7
9	9
11	10

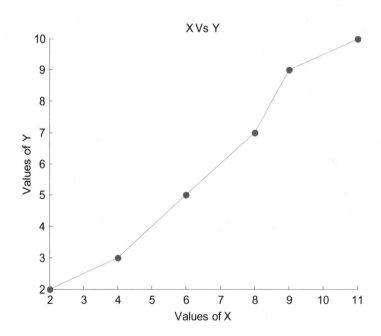

Fig. 4.1 Plot of example dataset

$$Y = \beta_0 + \beta_1 \times X. \tag{4.1}$$

In the above equation, input variable (X) is multiplied with a coefficient β_1 (beta one) and β_0 is bias coefficient. This equation is similar to equation of line $y = mx + c$. In case of multiple regression, there are more than one input variables, and this line becomes plane. Various methods are available to find out the values of regression coefficients; some of them are simple linear regression, ordinary least square, gradient descent, and regularized linear regression. In later sections, simple linear regression is explained with the above given dataset. This method requires very simple calculation of values like means, standard deviation, correlation, and covariance [1].

To correlate the regression analysis and its application, a simple example is prediction of weight of a person by the height, as given in Eq. (4.2),

$$\text{Weight} = \beta_0 + \beta_1 \times \text{Height}. \tag{4.2}$$

In the above equation, β_0 is bias coefficient and β_1 is coefficient of height. For instance, just assume that $\beta_0 = 0.2$ and $\beta_1 = 0.6$. These coefficient values are obtained from some method, which is described in later section. Now, assume the height of a person is 173 cm, and to predict weight, simply put the value of height in the above equation as, weight $= 0.1 + 0.4*173$. So, the result is 69.4, the unit may be considered as kilogram. As such, there is no role of units.

Simple linear regression: Implementation of simple linear regression is done by two simple and easy equations as given in Eqs. (4.3) and (4.4),

$$\beta_1 = \frac{\sum_{i=1}^{n}(X_i - \text{mean}(X)) \times (Y_i - \text{mean}(Y))}{\sum_{i=1}^{n}(X_i - \text{mean}(X))^2}, \tag{4.3}$$

$$\beta_0 = \text{mean}(Y) - \beta_1 \times \text{mean}(X). \tag{4.4}$$

Evaluate the above equations on the dataset given in Table 4.1, to get the values of β_1 and β_0. The values of β_1 and β_0 are 0.9578 and -0.3855, respectively. Now, these values act as model for coefficients or parameters for the given data. For predicting values through this mode, equation $P = -0.3855 + 0.9578 * X$ is used, where P represents the predicted values. X may be used for predicting the actual values of Y. Table 4.2 lists the X (input), P (predicted output), and Y (actual output).

Table 4.2 Sample of an example dataset	X (input)	P (predicted)	Y (output)
	2	1.5301	2
	4	3.4458	3
	6	5.3614	5
	8	7.2771	7
	9	8.2349	9
	11	0.1506	10

To estimate error in prediction, the simple measure is root mean square error (*RMSE*) as given in Eq. (4.5).

$$\text{RMSE} = \sqrt{\frac{\sum_{i=1}^{n}(P_i - Y_i)^2}{n}}. \tag{4.5}$$

RMSE value for the above model is 0.4537. This can be improved by using other advance method for regression modeling.

4.2.2   Linear Regression Using Gradient Descent

The calculation of coefficients of the model can also be done by gradient descent method. This is the process of minimizing a function by following the gradients of the objective function. This objective function is simply a mean square error or any function of error that is to be minimized. In this method, evaluation and updation of coefficient in each iteration have to be done for the minimization of error. The coefficients of model are also known as weight vectors ($W = [\beta_0 \beta_1]$). Simplified version of equation for updation of these weights is given in Eq. (4.6) [1]. The concept and derivation behind these equations are discussed in later sections of the chapter.

$$W = W - \alpha \times e \times X, \tag{4.6}$$

where α is the learning rate, e is the error, and X is the input.

To provide a very basic understanding of this method, few iterations of this gradient-based approach are demonstrated below:

Initially, $\beta_0 = 0$ and $\beta_1 = 0$ and the value of $\alpha = 0.01$, and the following steps are repeated in each iteration.

Steps of gradient descent approach
$P_i = \beta_0 + \beta_1 \times X_i$
$e_i = P_i - Y_i$
$\beta_0(t + 1) = \beta_0(t) - \alpha \times e_i$
$\beta_1(t + 1) = \beta_1(t) - \alpha \times e_i \times X_i$

In the above steps, i is the sample index of dataset and t is the current iteration. Calculation in iteration 1 is given below:

Iteration: 1
$P = 0 + 0*2 = 0$
$\Delta = 0 - 2 = -2$
$\beta_0(t + 1) = 0 - 0.01 \times (-2)$
$\beta_1(t + 1) = 0 - 0.01 \times (-2) \times 2$
After the first iteration: $\beta_0 = 0.02$ and $\beta_1 = 0.04$

Calculation in iteration 2 is given below:

Iteration: 2
$P = 0.02 + 0.04 * 4 = 0.18$
$\Delta = 0.18 - 3 = -2.82$
$\beta_0(t + 1) = 0.02 - 0.01 \times (-2.82)$
$\beta_1(t + 1) = 0.04 - 0.01 \times (-2.82) \times 4$
After the second iteration: $\beta_0 = 0.0482$ and $\beta_1 = 0.1528$

In the same manner, this calculation can also be performed for the third and the fourth iterations. MatLab code for the above method is given with the same dataset [2, 3]. This can be executed to cross check your calculation. To process all the samples, six iterations are required. After six iterations, the value of β_0 and β_1 are 0.148353 and 0.887801, respectively, with RMSE value of 2.8671. Now, the same process is repeated multiple times over the same dataset, known as epochs. One epoch means one complete processing of all the samples. So, multiple epochs are executed for more error reduction and better learning of the model.

```
regression_gradient_SLR.m

clc;
clear all;
close all;

% Gradient Descent Simple Linear Regression

X= [2 4 6 8 9 11];
Y= [2 3 5 7 9 10];

figure
hold
scatter(X,Y,'r','filled');
plot(X,Y);
title('X Vs Y')
xlabel('Values of X');
ylabel('Values of Y');

% Y = B0 + B1*X

B0=0;
B1=0;
alpha=0.001;
% Predicted Values (P) of Y
P=[];
error=[];
fprintf('\n iteration  B0    B1    error    P');

epoc=30;
c=0;
```

```
for j=1:epoc
for i=1:length(X)
P(i) = B0 + B1*X(i);
error(i+c) = P(i) - Y(i);
% Weight update
B0 = B0 - alpha * error(i+c);
B1 = B1 - alpha * error(i+c) * X(i);
fprintf('\n%d  %f %f %f %f',i+c,B0, B1, error(i+c),
P(i));
end

% Estimate Error
% Root Mean Square Error
figure
plot(error.^2);
title('Iteration Vs Square Error')
xlabel('Iterations');
ylabel('Square Error');

RMSE= sqrt(sum((P-Y).^2)/length(X))
```

Put the value of epoch = 5 in the above code to make 30 iterations. After 30 iterations, RMSE = 0.6453 and the plot of square error is shown in Fig. 4.2.

In the above code, by changing the dataset, the value of learning rate, and the number of epoch, different results will be achieved.

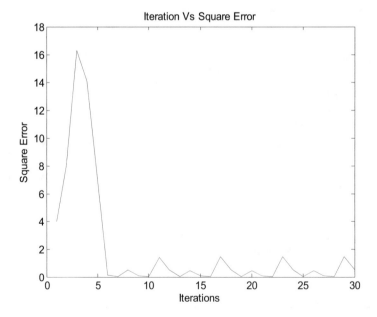

Fig. 4.2 Square error plot of 30 iterations

4.2.3   *Multiple Linear Regression*

Simpler to simple linear regression, multiple linear regression can be implanted following the same principle. In case of multiple linear regression, multiple input variables are used to build the relation by single output variable. A simplified equation for multiple linear regression is given in Eq. (4.7),

$$Y = \beta_0 + \beta_1 X_1 + \beta_2 X_2 + \cdots + \beta_q X_q, \tag{4.7}$$

where β_0 is the bias, and β_1 to β_q are the coefficients of input variables. The simple steps are discussed below for the implementation of multiple linear regression (4.1).

Steps of gradient descent approach for multiple linear regression

$$P_i = \beta_0 + \beta_1 \times X_i$$
$$e_i = P_i - Y_i$$
$$\beta_0(t+1) = \beta_0(t) - \alpha \times e_i$$
$$\beta_1(t+1) = \beta_1(t) - \alpha \times e_i \times X_1$$
$$\beta_2(t+1) = \beta_2(t) - \alpha \times e_i \times X_2$$
$$\vdots$$
$$\beta_q(t+1) = \beta_q(t) - \alpha \times e_i \times X_q$$

Now, for the implementation of the above steps, load MatLab built-in dataset carsmall.mat. It consists of acceleration, cylinders, displacement, horsepower, miles per gallon (MPG), model year, origin, and weight. Table 4.3 lists ten samples of horsepower, weight, and acceleration values from this dataset. For multiple regression, assume horsepower as $X1$, weight as $X2$, and acceleration as $Y1$. Figure 4.3 shows the scatter plot of the above dataset.

Normalization/Standardization: In general, normalization and standardization of data are the methods, by which raw data is processed. By these methods, a type of balance in values is created, if the range of value of the input data variables is very

Table 4.3 Car-small dataset for multiple linear regression

Horsepower ($X1$) (input)	Weight ($X2$) (input)	Acceleration (Y) (output)
130	3504	12
165	3693	11.50
150	3436	11
150	3433	12
140	3449	10.50
198	4341	10
220	4354	9
215	4312	8.50
225	4425	10
190	3850	8.50

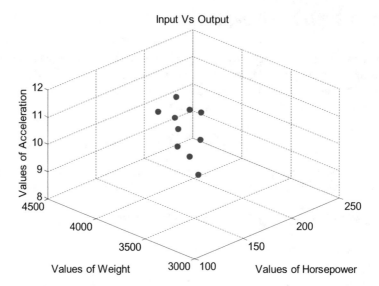

Fig. 4.3 Scatter plot of data

different from each other [4]. Some of the very commonly used techniques for these methods are:

- Rescaling data values between 0 and 1
- Transforming data values using a z-score or t-score.

The second technique is also called as standardization by z-score, and the data points will be standardized by Eq. (4.8).

$$x_i = \frac{x_i - \mu}{standared\ deviation},\tag{4.8}$$

where x_i is the sample value and μ is the mean value of all samples. The MatLab code for the above equation is given below, and the same has been used for data standardization.

---

standardization.m

```
function [Xnor, m, stddev] = standardization(X)
Xnor = [];
m = [];
stddev = [];
% Calculates mean and std dev of all input variables
for i=1:size(X,2)
    m(1,i) = mean(X(:,i));
    stddev(1,i) = std(X(:,i));
    Xnor(:,i) = (X(:,i)-m(1,i))/stddev(1,i);
end
```

After standardization of input data, below given code can be used for the implementation of multiple linear regression based gradient descent approach.

---

regression_gradient_MLR.m

```
clc;
clear all;
close all;

% Gradient Descent Multiple Linear Regression
load carsmall; %matlab built-in dataset

Xorg=[Horsepower(1:10) Weight(1:10)]; % taking two
input variables and 10 samples
X=standardization(Xorg); %Normalize the values of
input variables
Y=Acceleration(1:10); %one output variable and 10
samples

figure
scatter3(Xorg(:,1),Xorg(:,2),Y,'filled');
title('Input Vs Output')
xlabel('Values of Horsepower');
ylabel('Values of Weight');
zlabel('Values of Acceleration');

% Y = B0 + B1*X + B2*X

B0=0;
B1=0;
B2=0;

alpha=0.05;

% Predicted Values (P) of Y
P=[];
error=[];
fprintf('\n iteration  B0   B1  B2 error   P');

epoc=20;
c=0;
```

```
for j=1:epoc
for i=1:length(X)
P(i) = B0 + (B1*X(i,1)) + (B2*X(i,2));
error(i+c) = P(i) - Y(i);
% Weight update
B0 = B0 - alpha * error(i+c);
B1 = B1 - alpha * error(i+c)* X(i,1);
B2 = B2 - alpha * error(i+c)* X(i,2);

fprintf('\n%d %f %f %f %f %f ',i+c, B0, B1, B2,
error(i+c), P(i));
end
c=c+i;
end

% Estimate Error
% Root Mean Square Error
figure
plot(error.^2);
title('Iteration Vs Square Error')
xlabel('Iterations');
ylabel('Square Error');

RMSE= sqrt(sum((P'-Y).^2)/length(X))
```

After execution of the above code for one epoch, the error plot can be obtained as shown in Fig. 4.4, with RMSE value 10.0117. The updated values of coefficients β_0, β_1, and β_2 after one epoc are 0.9927, -0.0734, and -0.0569, respectively.

Now, by simply changing the value of alpha and number of epochs, the performance of regression model can be improved. The error plot after 20 epochs (200 iterations) of training is shown in Fig. 4.5. After 20 epochs, the RMSE value is improved, and now its value is 1.6825.

Further, by changing the alpha value to 0.05 and keeping the same number of epochs (20), the model is further improved, as shown by the error plot shown in Fig. 4.6. Now the value of RMSE is 0.8374. The final values of coefficients β_0, β_1, and β_2 are 10.2441, -1.0311, and 0.0902, respectively. Now, the same code can be used with different dataset and different values of alpha and epochs to train or fit the model.

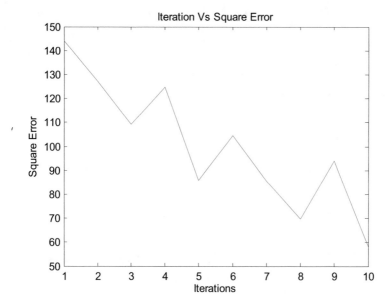

Fig. 4.4   Error plot of 1 epoch

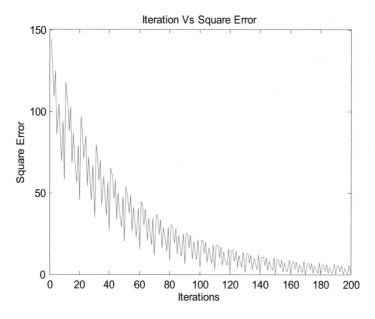

Fig. 4.5   Error plot of 20 epochs

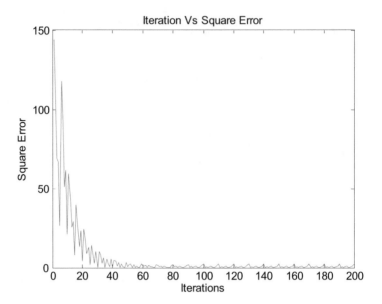

Fig. 4.6 Error plot of 20 epochs and alpha 0.05

4.3 Classification

4.3.1 Logistic Regression

Now, to move from prediction to classification model, the simplest and easy method is logistic regression. In this, binary classification (0 and 1) is performed for two classes. This can also be used for multiclass classification by making some modifications in the logistic regression, but for beginning, simple binary classification through logistic regression is explained. Logistic function used in this regression is also called as sigmoid function having S shaped curve. In logistic regression, a thresholding function is also used to convert the predicted value into class labels (binary 0 or 1) [1–3]. Figure 4.7 provides a simplified diagram of logistic regression model. As per this diagram, there are three main components: net sum, sigmoid function, and thresholding function. Figure 4.8 shows the visualization of linear model and logistic model, and by this, the concept of logistic regression becomes more clear.

The calculation of net sum function is the same in both the models. In case of logistic model, the net sum result (output) is passed into logistic function (sigmoid function), and the model will give predicted (P) value. The error is calculated between the class labels and predicted value. This error is utilized to update the

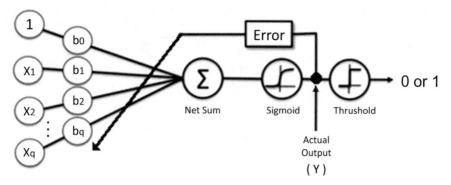

Fig. 4.7 Simplified diagram of logistic regression model

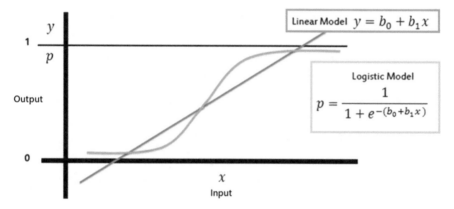

Fig. 4.8 Visualization of linear model and logistic model

values of coefficients (weights) of input variables. Here, generalized delta learning rule is used for weights updation, as discussed in later section. Once the training or learning process is completed by satisfying the permissible error or by number of iterations, final results are obtained by adding thresholding function after logistic function. This thresholding function will convert the predicted values to class labels, and finally accuracy or model can be calculated.

The above-discussed concept is implemented in MatLab code. To check the above concept and code, the dataset given in Table 4.4 is used. The same data is also plotted in Fig. 4.9. This code can be executed with different values of

Table 4.4 Dataset for logistic regression

X1 (input)	X2 (input)	Y (output)
3.20	1.70	0
2.10	3.10	0
2.40	2.30	0
1.60	2.10	0
2.90	1.90	0
6.80	0.60	1
7.10	2.11	1
5.50	1.90	1
5.10	3.30	1
8.30	2.40	1

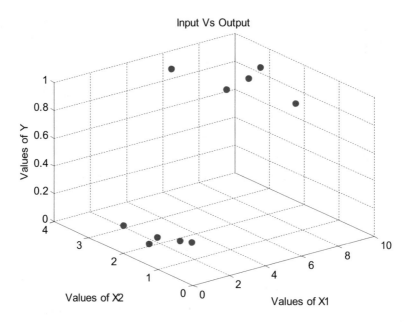

Fig. 4.9 Scatter plot of dataset for logistic regression

epoch and alpha. Figures 4.10, 4.11, and 4.12 show the error plot for different trials with different values of alpha and epoch as listed in Table 4.5 as summary of executions.

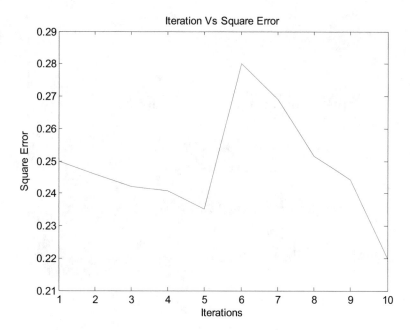

Fig. 4.10 Error plot for one epoch

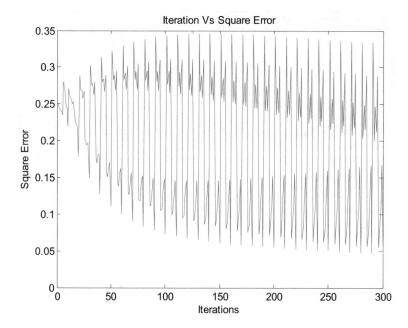

Fig. 4.11 Error plot for 30 epochs

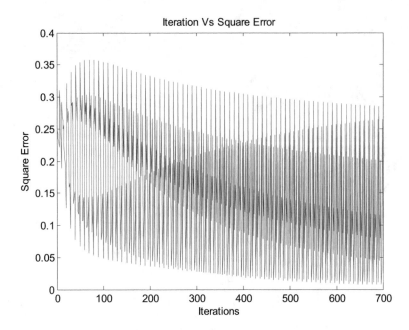

Fig. 4.12 Error plot for 70 epochs with alpha value 0.02

```
logistic_regression_gradient.m

clc;
clear all;
close all;

% Gradient Descent Logistic Regression
load data1;

X=data1(:,1:2); % taking two input variables and 10
samples
Y=data1(:,3); %one output variable and 10 samples

figure
scatter3(X(:,1),X(:,2),Y,'filled');
title('Input Vs Output')
xlabel('Values of X1');
ylabel('Values of X2');
zlabel('Values of Y');

% output = B0 + B1*X + B2*X
% P(class=0)=1/(1+(e^(-output)))

B0=0;
B1=0;
B2=0;
```

Table 4.5 Summary of different trials with different number of epochs and alpha value

S. no	Epoch	Alpha	β_0	β_1	β_2	RMSE	Accuracy
1	1	0.01	0.000112	0.026010	−0.000890	0.4978	50
2	30	0.01	−0.084512	0.224504	−0.193135	0.4163	70
3	70	0.02	−0.345239	0.484511	−0.630043	0.3337	90

```
alpha=0.01;

% Predicted Values (P) of Y
P=[];
error=[];
fprintf('\n iteration  B0   B1   B2 error   P');

epoc=1;
c=0;

for j=1:epoc
for i=1:length(X)
op(i)= B0 + (B1*X(i,1)) + (B2*X(i,2));
P(i)=1/(1+exp(-op(i)));
error(i+c) = P(i) - Y(i);
% Weight update
B0 = B0 + alpha * (Y(i)-P(i))*P(i)*(1-P(i))*1;
B1 = B1 + alpha * (Y(i)-P(i))*P(i)*(1-P(i))*
X(i,1);
B2 = B2 + alpha * (Y(i)-P(i))*P(i)*(1-P(i))*
X(i,2);
fprintf('\n%d %f %f %f %f %f ',i+c, B0, B1, B2,
error(i+c), P(i));
end
c=c+i;
end

% Estimate Error
% Root Mean Square Error
figure
plot(error.^2);
title('Iteration Vs Square Error')
xlabel('Iterations');
ylabel('Square Error');

RMSE= sqrt(sum((P'-Y).^2)/length(X))


% Converting Predections to Crisp Values (by
thresholding)
% If(P < 0.5) then class is 0, otherwise class is 1
```

```
for i=1:length(X)
op(i)= B0 + (B1*X(i,1)) + (B2*X(i,2));
P(i)=1/(1+exp(-op(i)));

if (P(i)<0.5) % Thrushold
    P(i)=0;
else
    P(i)=1;
end
end

% Accuracy Calcualtion
% Accuracy = (Correct predection/ Total Prediction)
x 100
CP=0;
TP=length(Y);
for i=1:length(Y)
if (P(i)==Y(i))
    CP=CP+1;
end
end

Acc=(CP/TP)*100
```

It is very clear from Table 4.5 that by executing more iterations (number of epochs), the error will be reduced and the accuracy of the model will improve. Further by changing the value of alpha, more improvement in model can be done.

In later section, fundamentals of ANN are discussed. It is well known that very little modifications are required to convert the logistic regression into ANN, and inside the ANN, multiple regression units work together. ANN is just a combination of multiple regression units.

4.3.2 ECG Classification Using Logistic Regression

A very basic example of classification problem in taken up here for better understanding of feature extraction and classification processes.

Dataset specifications: This dataset is taken from MendeleyECG signals (1000 fragments), DOI: 10.17632/7dybx7wyfn.3 [5], and it consists of ECG recording collected through PhysioNet service (http://www.physionet.org) from the MIT-BIH Arrhythmia database. In full dataset, there are 17 classes for different conditions of heart. Only two classes are considered here to demonstrate the logistic regression. normal sinus rhythm as class 1 and atrial premature beats as class 2. All ECG signals were recorded at a sampling frequency of 360 Hz and a gain of 200 mV. Only first

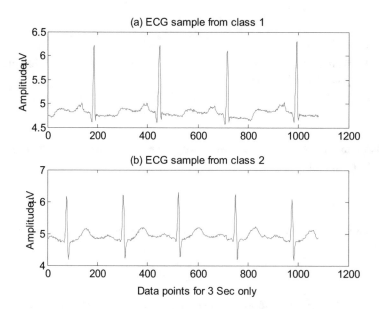

Fig. 4.13 Plot of (**a**) ECG sample of class 1 and (**b**) ECG sample of class 2

3 s data is taken here and is divided by gain to get the actual amplitude. Ten samples of each class is taken as dataset, first ten samples are of normal condition (normal sinus rhythm), and the last ten samples are of abnormal condition (atrial premature beats). Their corresponding class labels are 0 and 1 for class 1 and class 2, respectively. Figure 4.13 shows the ECG sample of class 1 and class 2. It can be noticed that the portion of ECG wave just after the peak is different in these samples.

4.3.2.1 ECG Feature Extraction

The main aim of feature extraction is to extract some useful information from the data. This information is able to differentiate different data samples; it may be signals or images. This will also lead to data size reduction. Simple time domain features are explained here. In case of ECG signals, 3 s data (360*3 = 1080 samples) is passed into MatLab program to extract features based on the equations given below for simple time domain statistical features [4].

Mean: It is denoted as μ, is the average of all the voltage level of signal, and is given by Eq. (4.9),

$$\mu = \frac{\sum_{i=1}^{N} x_i}{N}.$$

(4.9)

Variance (VAR): It measures the signal values distribution from mean of the signal, as given by Eq. (4.10),

$$VAR = \frac{1}{N} \sum_{i=1}^{N} (x_i - \mu)^2. \tag{4.10}$$

Root mean square (RMS): It is square root of the average squared sum of signal values as given by Eq. (4.11),

$$x_{rms} = \sqrt{\frac{1}{N} \sum_{i=1}^{N} x_i^2}. \tag{4.11}$$

Shape factor: It is RMS value divided by the mean of absolute value of the signal as given by Eq. (4.12),

$$x_{sf} = \frac{x_{rms}}{\frac{1}{N} \sum_{i=1}^{N} |x_i|}. \tag{4.12}$$

Simple sign integral (SSI): It determines the energy of signal and is obtained by Eq. (4.13),

$$SSI = \sum_{i=1}^{N} x_i^2. \tag{4.13}$$

For the above given features, a simple MatLab script is created, in which any type of one-dimensional signal is passed to get back the feature values (f1–f5).

```
feature_ext.m
function [mu v rms sf ssi]=feature_ext(x)
N=length(x);

% f1 Mean
mu = sum(x)/N;

% f2 Variance
v = sum(((x)- mu).^2)/N;

% f3 Root Mean Square
rms = sqrt(sum(x.^2)/N);

% f4 Shape factor
sf = rms/(sum(abs(x))/N);

% f5 Simple Sign Integral
ssi = sum(x.^2);

end
```

4.3.2.2 ECG Classification

After obtaining values of features, these features are treated as input to logistic regression model, and the output is desired as class labels. Same process of logistic regression has been followed as described above, just by changing the dataset as values of features. As per the above feature extraction part, there are totally five features, only two features are considered at a time. Through the MatLab code given below, the entire process of feature extraction and classification can be easily demonstrated.

```
ecg_classification.m

clc;
clear all;
close all;

% ECG classification by Logistic Regression
load nor_ecg; % 10 samples of normal ECG
load apb_ecg; % 10 samples of abnormal ECG (Atrial
premature beat)

figure
subplot(2,1,1)
plot(nor_ecg(1,1:360*3));
title('(a) ECG sample from class 1');
ylabel('Amplitude \muV');
subplot(2,1,2)
plot(apb_ecg(1,1:360*3));
title('(b) ECG sample from class 2');
ylabel('Amplitude \muV');
xlabel('Data points for 3 Sec only')


% Feature extration
fx_nor=[];
for i=1:10
[mu v rms sf ssi]=feature_ext(nor_ecg(i,1:360*3)); %
taking only 3 sec ECG data (sampling freq is 360Hz)
fx_nor(i,:)=[mu v rms sf ssi];
end

fx_apb=[];
for i=1:10
[mu v rms sf ssi]=feature_ext(apb_ecg(i,1:360*3)); %
taking only 3 sec ECG data (sampling freq is 360Hz)
fx_apb(i,:)=[mu v rms sf ssi];
end

input_data=[fx_nor; fx_apb];
output_class_label=[zeros(1,10) ones(1,10)];

Xnor = standardization(input_data);
X=Xnor(:,[1,2]); % taking two input variables and 10
```

```
samples
Y=output_class_label';


figure
scatter3(X(:,1),X(:,2),Y,'filled');
title('Input Vs Output')
xlabel('Values of X1');
ylabel('Values of X2');
zlabel('Values of Y');


% output = B0 + B1*X + B2*X
% P(class=0)=1/(1+(e^(-output)))


B0=0;
B1=0;
B2=0;


alpha=0.8;

% Predicted Values (P) of Y
P=[];
error=[];
fprintf('\n iteration  B0   B1   B2 error   P');


epoc=20;
c=0;

for j=1:epoc
for i=1:length(X)
op(i)= B0 + (B1*X(i,1)) + (B2*X(i,2));
P(i)=1/(1+exp(-op(i)));
error(i+c) = P(i) - Y(i);
% Weight update
B0 = B0 + alpha * (Y(i)-P(i))*P(i)*(1-P(i))*1;
B1 = B1 + alpha * (Y(i)-P(i))*P(i)*(1-P(i))* X(i,1);
B2 = B2 + alpha * (Y(i)-P(i))*P(i)*(1-P(i))* X(i,2);

fprintf('\n%d %f %f %f %f %f ',i+c, B0, B1, B2,
error(i+c), P(i));
end
c=c+i;
```

```
end

% Estimate Error
% Root Mean Square Error
figure
plot(error.^2);
title('Iteration Vs Square Error')
xlabel('Iterations');
ylabel('Square Error');

RMSE= sqrt(sum((P'-Y).^2)/length(X))



% Converting Predections to Crisp Values (by
Thrusholding)
% If(P < 0.5) then class is 0, otherwise class is 1

for i=1:length(X)
op(i)= B0 + (B1*X(i,1)) + (B2*X(i,2));
P(i)=1/(1+exp(-op(i)));

if (P(i)<0.5) % Thrushold
    P(i)=0;
else
    P(i)=1;
end
end

% Accuracy Calcualtion
% Accuracy = (Correct predection/ Total Prediction) x
100
CP=0;
TP=length(Y);
for i=1:length(Y)
if (P(i)==Y(i))
    CP=CP+1;
end
end

Acc=(CP/TP)*100
```

After executing different trial of the above code with different values of parameters like alpha and number of epochs, different observations were noticed, out of which one is shown in the form of error plot in Fig. 4.14. Summary for ECG classification for different trials with different pairs of features is given in Table 4.6.

In the same manner, performance of the model can be analyzed by changing the data signals, features, number of features, value of alpha, and number of epochs. The code can be easily modified by adding more input variables in the model.

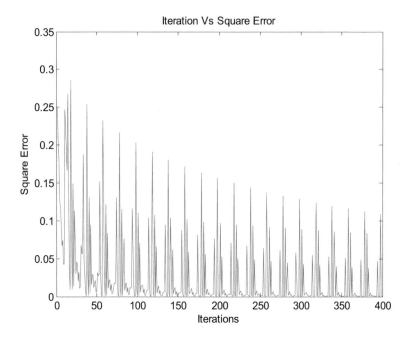

Fig. 4.14 Error plot for 20 epochs and alpha 0.8 for feature 1 and 2

Table 4.6 Summary of different trial for ECG classification with different pair of features

S. no	Feature no.	Epoch	Alpha	β_0	β_1	β_2	RMSE	Accuracy
1	1,4	10	0.8	0.465916	2.109560	2.690791	0.1171	100
2	1,3	20	0.8	0.321635	2.111929	2.178224	0.2171	95
3	1,2	20	0.8	0.548536	1.943690	2.754677	0.1203	100

4.4   Basics of Artificial Neural Networks (ANN)

The basic terminology used in ANN is explained in very simple way through the following terms [6–8]:

1. **Neurons/Node:** Neuron is a biological term that represents a basic element in the brain. It is the inspiration for modeling of a node. A node is a term used to represent a basic element in ANN, which mathematically works on the concept of regression. Generally, each node consists of weighted sum unit (also called as net sum) and activation function.
2. **Activation Functions:** It is a mathematical function that converts input value (the net sum value) through mathematical equation to output value. Sigmoid function is an example of activation function.

3. **Cost Function:** This term is taken from the optimization field; it is just an error function whose values are required to be minimized. Generally, it is calculated as difference between the predicted output and actual (desired) output of ANN.

4. **Gradient Descent Learning Algorithm:** This is already explained above; this is the approach by which the values of weights of ANN are updated by following the gradient of error function. This utilizes generalized delta learning rule as discussed below in detail.

5. **Back Propagation:** It is a term used while discussing the learning of ANN. More specifically, it is known as "error back propagation". This means error is propagated in backward direction to update the weights of ANN. While input data is propagated in forward direction, it is more specifically called as "feed forward network."

4.4.1 ANN Architecture and Its Training

The structure of node is the basic element of ANN. Multiple nodes create the ANN. Figure 4.15 shows the structures of a neuron and a node. In neuron, dendrites collects the multiple input signals, nucleus sum up all the input signal, and axon passed output signal, only if it exceeds certain level. This working of neuron is modeled as a

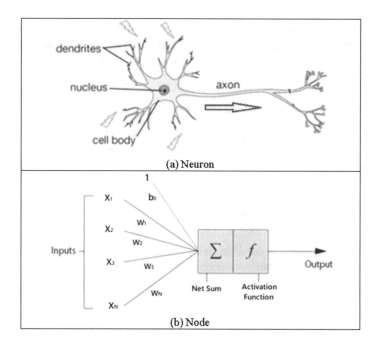

Fig. 4.15 Structure of (**a**) neuron and (**b**) node

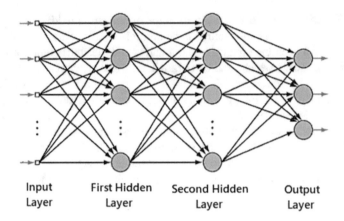

Input   First Hidden   Second Hidden   Output
Layer   Layer   Layer   Layer

Fig. 4.16 ANN architecture

node, where net sum unit processes the input data, and activation function acts as axon.

A simple node is formulated as Eq. (4.14)

$$f(x) = b_0 + \sum_{i=1}^{N} X_i \times W_i, \qquad (4.14)$$

where X_i is the input data sample and W_i is weights (coefficients) of the node. Bias is represented through b_0.

Basically, ANN consists of three types of layers: input layer, hidden layer, and output layer. Each of these layers consists of multiple nodes. Each node in hidden layers takes input from the output of its previous layer. Input layer simply passes the input data as it is in the ANN, while the output layer gives the final output as result. Figure 4.16 shows the architecture of ANN.

Following steps need to be performed to train the ANN:

Step 1: *Initialize weights (may initialized with zero or some random values).*
Step 2: *Calculate output of ANN and also calculate error.*
Step 3: *Adjust/update weight to reduce error by learning rule. Generally, delta learning rule is used.*
Step 4: *Repeat steps 2 and 3.*
Step 5: *Repeat steps 2–4 until error reduces to acceptable level (one complete execution of steps 2–4 is called as epoch).*

The difference between the delta and generalized delta rules used for learning is discussed below. The equation of delta rule is $W_{ij} = W_{ij} + \alpha \times e_i \times X_i$, while in generalized delta rule, e_i is replaced by δ_i as given in Eq. (4.15).

$$W_{ij} = W_{ij} + \alpha \times \delta_i \times X_i. \tag{4.15}$$

Here, δ_i is the derivative of activation function of node i multiplied by error as in Eq. (4.16).

$$\delta_i = f'(\text{Activation function}) \times e_i. \tag{4.16}$$

For linear activation function, the derivative is 1. While, the derivative of sigmoid function $f(x) = 1/(1 + x^{-1})$ is $f'(x) = f(x) \times (1 - f(x))$.

Therefore, final equation of δ_i is as given in Eq. (4.17),

$$\delta_i = f(x) \times (1 - f(x)) \times e_i \tag{4.17}$$

and the complete updated equation is given in Eq. (4.18)

$$W_{ij} = W_{ij} + \alpha \times f(x) \times (1 - f(x)) \times e_i \times X_i. \tag{4.18}$$

Sometimes, it is also used as $W_{ij} = W_{ij} + \Delta W_{ij}$, where, $\Delta W_{ij} = \alpha \times f(x) \times (1 - f(x)) \times e_i \times X_i$.

Now, to update the weights or for training of ANN, there are three possible ways. The first one is stochastic gradient descent (SDG) approach, the second is Batch approach, and the last one is Mini Batch approach.

SGD: Weights are updated in each iteration means after processing of each sample.

Batch: Error is calculated for all data samples, and ΔW_{ij} is also calculated for all samples, but average of all ΔW_{ij} is used to update the weights. In this, $\Delta W_{ij} = \frac{1}{N} \sum_{k=1}^{N} (\Delta W_{ij}(k))$ is used, where, N is the total number of samples. It is stable learning as compared to SGD.

Mini Batch: It is the combination of SGD and Batch. In this, all samples are divided into small batches, and after each batch, weights are updated. For example, total 30 samples are there in database and batch size is 3, so after processing three samples, weights are updated once. Hence, weights are updated only ten times.

4.4.2 Activation Functions

Activation functions play an important role in ANN. For improvement in the performance of ANN, different activation functions can be tried after net sum unit. Some of the famous activation functions are explained below with their equation and output responses [8].

1. **Step function**

Equation:	$f(x) = \begin{cases} 0 & \text{if } x < 0 \\ 1 & \text{if } x \geq 0 \end{cases}$
Response:	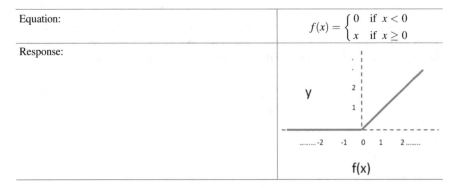

2. **Sigmoid function**

Equation:	$f(x) = \left\{ \frac{1}{1+e^{-(x)}} \right.$
Response:	

3. **Hyperbolic tangent function**

Equation:	$f(x) = \left\{ \frac{e^x - e^{-x}}{e^x + e^{-x}} \right.$
Response:	

4. **Rectified linear unit (ReLU) function**

Equation:	$f(x) = \begin{cases} 0 & \text{if } x < 0 \\ x & \text{if } x \geq 0 \end{cases}$
Response:	

4.4.3   Multiclass Classification

In case of multiclass classification problem, multiple output neurons are used. For example, in database, the classes are given as $0 =$ first class, $1 =$ second class, and $2 =$ third class. These classes can be encoded in ANN as $0=[1\ 0\ 0]$, $1=[0\ 1\ 0]$, and $2=[0\ 0\ 1]$, and three neurons are used in the output layer. The final output as class label is calculated by average max function, and this will return the index of the output vector with the highest value as the class label.

4.5   Some Useful Tips

Some very basic things that are considered before starting any classification or prediction study are as following:

Noise reduction: Identify samples which are not following statistical properties of dataset, also known as outliers, if possible remove them.

Collinearity: Identify the pairs of input data variables, which are highly correlated, remove one variable from each of the most correlated pair.

Rescale: This basically means the normalization of variables, if mean and standard deviation of variables is large, then it affects the model performance, rescale it.

References

1. J. Brownlee, Master Machine Learning Algorithms: discover how they work and implement them from scratch. Machine Learning Mastery (2016), https://machinelearningmastery.com/master-machine-learning-algorithms/
2. https://in.mathworks.com/help/stats/linearmodel.feval.html
3. https://in.mathworks.com/help/stats/regress.html
4. A.G. Bluman, *Elementary statistics: A step by step approach* (McGraw-Hill Higher Education, New York, NY, 2009)
5. P. Pławiak, ECG signals (1000 fragments), Mendeley Data, V3 2017, https://doi.org/10.17632/7dybx7wyfn.3
6. S.N. Sivanandam, S.N. Deepa, *Introduction to neural networks using Matlab 6.0* (Tata McGraw-Hill Education, New York, 2006)
7. I.A. Basheer, M. Hajmeer, Artificial neural networks: fundamentals, computing, design, and application. J. Microbiol. Methods **43**(1), 3–31 (2000)
8. N.P. Padhy, S.P. Simon, *Soft computing: with MATLAB programming* (Oxford University Press, Oxford, 2015)

Chapter 5
Basics of Matlab Programming

5.1 MatLab Programming Basics

MatLab is normally the first choice for scientific computations. MatLab programming is very simple to learn, and the beginners are advised to visualize their data and problem as matrix [1]. By this visualization, all the concepts of MatLab programming become very easy to understand.

There are some important windows inside MatLab GUI, in default view, use and work of each window is described below:

Current directory: This is the window from where access of files and folders is easy. The locations shown in this window is called present working directory. MatLab program files under present working directory are directly accessed and executed.

Command window: This is the main window of MatLab, from where all the commands and MatLab program file (scripts) are executed. Results are also displayed on this window.

Workspace: This is the place, where the entire variables are visible with their size, type, and values. This window can be further explored with each variable as Variable Editor Window.

Command history: This is the window, where all the previous commands are visible. It is the type of log record of previous commands executed.

After getting familiar with MatLab GUI, calculations in command window are explained in the following section [1, 2].

© The Author(s) 2021
M. K. Ahirwal et al., *Computational Intelligence and Biomedical Signal Processing*,
SpringerBriefs in Electrical and Computer Engineering,
https://doi.org/10.1007/978-3-030-67098-6_5

5.1.1   *Calculations in Command Window*

Examples and exercise of simple calculations with mathematical functions over constants and variables are given below [1–3] (double arrow ">>" symbol represents the command prompt terminal).

(a) Simple Calculation
      Calculations can be directly and easily performed on command prompt.

```
>> -4/(2.9+6.13)^3
```

ans =

```
 -0.0054
>> (2+5i)*(2-5i)
```

ans =

```
 29
>>cos(pi/4)
```

ans =

```
 0.7071
>>exp(tan(0.5))
```

ans =

```
 1.7269
```

(b) Use of Variables
      Calculations with variables and constants can also be directly and easily performed on command prompt as given below:

```
>> m=7;
>> n=9;
>>m+n
```

ans =

```
 16
>> r=pi/2    (pi is constant)
```

r =

```
 1.5708
>> s=sin(r)
```

s =

```
 1
```

(c) Floating Point Number Precision Control

When displaying the results of calculations or values of variables, "format" command is used to control the precision of numbers.

```
>>format short
>> 3.4
ans =
   3.4000    (only 4 decimals in the result)
>>format long
>> 3.4
ans =
   3.400000000000000   (15 decimals in the result)
```

5.1.2 Arithmetic Operators

Basic arithmetic operators are given in Table 5.1, and their precedence is also listed in Table 5.2.

(a) Example of Change in Precedence

Use of parentheses to change or control the precedence of operators is demonstrated by examples.

>> (1+2)*3	>> 1+2*3
ans =	ans =
9	7

Table 5.1 Basic arithmetic operators

Symbols	Operation	Example
+	Addition	4+5
-	Subtraction	4-5
*	Multiplication	4*5
/	Division	4/5
^	Power	3^2

Table 5.2 Precedence of operators

Precedence	Arithmetic operation
1	Contents inside the innermost parentheses are evaluated first, then the outermost parentheses are evaluated
2	In left to right manner, exponentials are evaluated in an equation
3	In left to right manner, multiplication and division are performed
4	In left to right manner, addition and subtraction are performed

(b) Exercise for Coding of Equation

Some examples of calculations and equations are given below, try to code them and check the obtained answer with the given answer.

Exercise 1:

$$\frac{1}{2+3^2} + \frac{4}{5} \times \frac{6}{7}$$

Correct answer = 0.7766

Exercise 2:

$$f = \left(x^2 + y^2\right)^4 - \left(x^2 - y^2\right)^2$$

Evaluate the above equation two times, first time for $x = 1$ and $y = 1$, second time for $x = -1$ and $y = -1$.

Correct answer: 16

If both the answers are 16, then the equation is correctly coded.

5.1.3 Managing the Workspace and Miscellaneous Commands

Some miscellaneous commands are given in Table 5.3. These commands will help to speed up the programming and also helps in managing the workspace.

To continue a line with equations or commands, use . . . (three dots).

Table 5.3 Miscellaneous commands

Commands	Operation
Clc	This will clear the command window
Clear	To empty workspace, delete all the variables
Who	To know variable names in workspace
Whos	To know variable names with size and memory consumed
Diary	Diary on, diary off, diary <filename>, this will create a text file of command history of command prompt
ctrl + c	To abort a MatLab computation
Help	Help <command>, explore about the given command
Lookfor	Lookfor <keyword>, search the given keyword and suggest matched commands

5.1.4   *Mathematical Functions and Predefined Constants*

Some of the frequently used mathematical functions and predefined constants are listed in Tables 5.4 and 5.5, respectively.

(a) Exercise for Coding of Expression
   Below is a mathematical expression with the value of variables and its result, code this expression and check the obtained answer with the given answer.

$$z = e^{-a} \sin(x) + 10\sqrt{y}$$

for $a = 5$, $x = 2$, and $y = 8$.
If answer $z = 28.2904$, then the equation is coded correctly.

5.1.5   *Working with Matrix*

(a) Creation of Matrix
   Different ways for creation of matrix are demonstrated with examples. Simplest way is initialization of matrix with values.

```
>> a = [1 2 3; 4 5 6; 7 8 9]
or
>> a = [1, 2, 3; 4, 5, 6; 7, 8, 9]
```

   Both creates the same matrix. While creating matrix, rows are separated by colon. This matrix can be visualized as:

Table 5.4 Mathematical functions

Syntax	Function	Syntax	Function
cos(x)	Cosine function	abs(x)	Absolute value
sin(x)	Sin function	max(x)	Maximum value
tan(x)	Tan function	min(x)	Minimum value
exp(x)	Exponential function	ceil(x)	Round toward upper value
sqrt(x)	Square root	conj(x)	Complex conjugate
log(x)	Natural logarithm	floor(x)	Round toward lower value
log10(x)	Common logarithm	round(x)	Round to nearest integers

Table 5.5 Predefined constant values

Pi	$\pi = 3.14159$
i,j	Imaginary unit (c = a + b*1i)
Inf	Infinity
NaN	Not a number

$$a = \begin{bmatrix} 1 & 2 & 3 \\ 4 & 5 & 6 \\ 7 & 8 & 9 \end{bmatrix}$$

Other ways are based on the use of functions like "zeros," "once," "eye," and "rand." In these functions, pass first parameters as value of row and second as column.

```
>>a=zeros(3,3)
a =
     0    0    0
     0    0    0
     0    0    0
>>a=ones(2,2)
a =
     1    1    1
     1    1    1
     1    1    1
>>a=eye(2,2)
a =
     1    0    0
     0    1    0
     0    0    1
>> a=rand(3,3)
a =
     0.8147   0.9134   0.2785
     0.9058   0.6324   0.5469
     0.1270   0.0975   0.9575
```

These functions create the square matrix, if only one value is passed. Some examples of one-dimensional arrays are given below:

```
>> a=zeros(1,3)
a =
     0    0    0
>> a=ones(1,4)
a =
     1    1    1    1
>> a=eye(1,5)
a =
     1    0    0    0    0
>> a=rand(1,6)
a =
     0.9649   0.1576   0.9706   0.9572   0.4854   0.8003
>> a=rand(3,1)
a =
     0.0357
     0.8491
     0.9340
```

```
>> a=zeros(2,1)
a =
    0
    0
```

(b) Accessing Matrix Elements

To access a specific element or group of elements, simple exercise with examples is given below:

```
>> y=[1 2 3; 4 5 6; 7 8 9]
y =
    1   2   3
    4   5   6
    7   8   9
```

With matrix name and index of row and column like "matrix(row, column)," elements are accessible. Some examples are given below.

```
>>y(2,3)
ans =
    6
>>y(1,1)
ans =
    1
```

By passing only one index, elements are returned in row major order.

```
>>y(2)
ans =
    4
>>y(8)
ans =
    6
>>y(5)
ans =
    5
```

To access dissentious indexes, use the following syntax:

```
>>y(3,[1,3])
ans =
    7   9
```

```
>>y([3,1],2)
ans =
    8
    2
```

Colon operator is used to get all the elements. How to use it is demonstrated below:

```
>>y(:,:)
ans =
    1   2   3
    4   5   6
    7   8   9
>>y(1,:)
ans =
    1   2   3
>>y(:,1)
ans =
    1
    4
    7
>>y(1:3)
ans =
    1   4   7
```

```
>>y(3:end)
ans =
    7   2   5   8   3   6   9
>>y(3,2:3)
ans =
    8   9
```

As an exercise, execute the following commands and observe their outputs.

```
>>y(end:-1:1,end)
>>y([1 3],[2 3])
>>y(3,:) = []
>>y = [y(1,:);y(2,:);[7 8 0]]
```

(c) Dimension of Matrix

When working with multidimensional matrix, "size" command will help to identify the size or length (number of elements) of each dimension. This command is very useful when working with signals, images, and other high dimensional data matrix.

```
>> y=[1 2 3; 4 5 6; 7 8 9]
>>size(y)
ans =
     3   3
>> [m,n]=size(y)
m =
    3
n =
    3
```

(d) Expression and Function-Based Matrix

Below is an example of expression and function-based matrix. Similar approach can be followed in any place, where the answer of the expression is to be directly stored in matrix.

$$c = \begin{bmatrix} x+2 & 3.4 * \sin(x) & \text{Inf} \\ x/3 & \log(1) & \sqrt{4} \\ 2/3 & \text{NaN} & x * \cos(0) \end{bmatrix}$$

Take the value of x = 1.

```
>> c=[x+2 3.4*sin(x) Inf; x/3 log(1) sqrt(4); 2/3 NaN x*cos(0)]
c =
   3.0000   2.8610     Inf
   0.3333        0  2.0000
   0.6667      NaN  1.0000
```

Now, try to code for the below example:

$$d = \begin{bmatrix} x*2 & 4 * \cos(x)/3 & \text{NaN} \\ \log(9) & \log(2)/2 & \sqrt{10} \\ (2+3)*3 & e^{-x} & \sqrt{x} \end{bmatrix}$$

Take the value of x = 1.
Correct answer is,

```
d =
    2.0000   0.7204     NaN
    2.1972   0.3466  3.1623
   15.0000   0.3679  1.0000
```

(e) Matrix Arithmetic

In Table 5.6, matrix arithmetic operations with operators are explained. Element-wise operations are also listed.

One can also try all the above examples from sections (a) to (e) on three-dimensional matrixes. Use "cat" command to create the three-dimensional matrixes. It simply concatenates two or more matrixes.

```
>> a=cat(3,[1 2; 3,4],[5 6; 7 8],[9 10; 11 12])
```

Table 5.6 Matrix arithmetic operations

Matrix with operator	Use of operator
X + Y	Valid only, if both are square matrix
X * Y	Valid only, if the number of column in X is equal to the number of rows in Y
X^2	Valid only, if X is a square matrix
X*α or α*X	Each element is multiplied by α (any constant value)
Element-wise operations	
X.*Y	To multiply the corresponding elements of both matrix
X./Y	To divide the elements of X by the corresponding elements of Y
X.^Y	To use elements of Y as power over the corresponding elements of X

5.1.6 Figure Plotting

In this section, different types of plots with their detailing are demonstrated. Generation of series of data values for making it as axis are also described with simple functions.

(a) Use of Plot Command

Consider variables x and y both as row matrix or column matrix of the same length. x may be assumed as time scale and y meaning full data. Different plots of x and y are shown in Fig. 5.1.

```
>> x = [1 2 3 4 5 6];
>> y = [3 -1 2 4 5 1];
>> plot(x)
>> plot(y)
>> plot(x,y)
>> plot(y,x)
```

Procedure to plot any function, for example, $y = 4x^2 + 2x - x$ between interval 0 and 10 is given below with the plot in Fig. 5.2.

```
>> x=[1:10]
x =
   1   2   3   4   5   6   7   8   9   10
>> y=4*(x.*x)+2*x-x
y =
   5   18   39   68   105   150   203   264   333   410
>>plot(x,y)
```

Procedure to plot the function sin(x) between interval [0; 2π] is given below with the plot in Fig. 5.3.

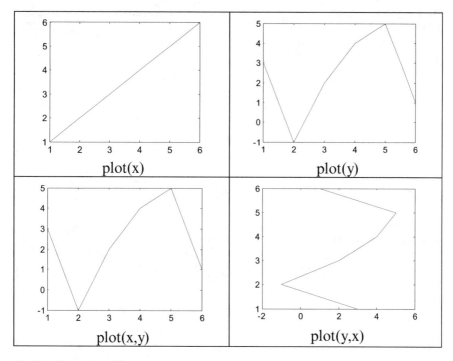

Fig. 5.1 Plots of *x* and *y*

Fig. 5.2 Plot of
$y = 4x^2 + 2x - x$

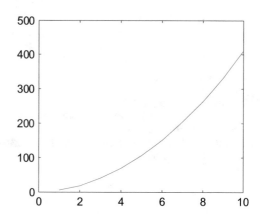

```
>> x = 0:pi/10:2*pi
x =
 Columns 1 through 7
      0   0.3142   0.6283   0.9425   1.2566   1.5708   1.8850
 Columns 8 through 14
   2.1991   2.5133   2.8274   3.1416   3.4558   3.7699   4.0841
 Columns 15 through 21
   4.3982   4.7124   5.0265   5.3407   5.6549   5.9690   6.2832
```

Fig. 5.3 Plot the function
sin(x)

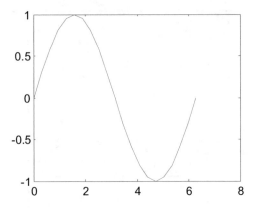

Fig. 5.4 Plot the function
sin(x) with labels and titles

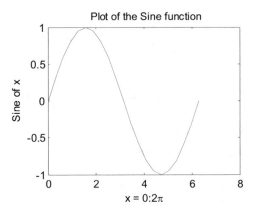

```
>> y = sin(x)
y =
 Columns 1 through 7
      0   0.3090   0.5878   0.8090   0.9511   1.0000   0.9511
 Columns 8 through 14
  0.8090   0.5878   0.3090   0.0000  -0.3090  -0.5878  -0.8090
 Columns 15 through 21
 -0.9511  -1.0000  -0.9511  -0.8090  -0.5878  -0.3090  -0.0000
>>plot(x,y)
```

Now, to add titles, axis labels, and annotations to above plot, following
commands are used (Fig. 5.4).

```
>>xlabel('x = 0:2\pi')
>>ylabel('Sine of x')
>>title('Plot of the Sine function')
```

To plot multiple data variables in single plot as shown in Fig. 5.5, following
lines of code can be used.

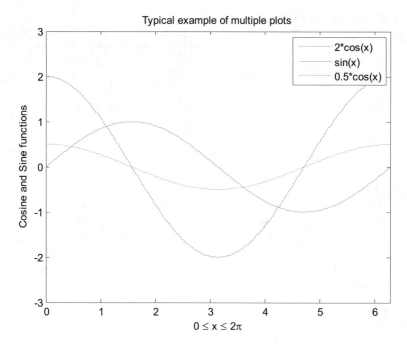

Fig. 5.5 Multiple sin function in a single plot

```
>> x = 0:pi/100:2*pi;
>> y1 = 2*cos(x);
>> y2 = sin(x);
>> y3 = 0.5*cos(x);
>>plot(x,y1,'--',x,y2,'-',x,y3,':')
>>xlabel('0 \leq x \leq 2\pi')
>>ylabel('Cosine and Sine functions')
>>legend('2*cos(x)','sin(x)','0.5*cos(x)')
>>title('Typical example of multiple plots')
>> axis([0 2*pi -3 3])
```

(b) Detailing of Plots by Symbols for Colors

To further explore more detailing of figure and plots, few symbols and markers are used as listed in Table 5.7. These symbols can be used to change the colors and shape/symbols of plots.

5.1.7 Steps in MatLab Programming

Above you learned about the basics of command line execution of in-built functions of MatLab and basic commands for simple expression. Below are the steps to prepare a MatLab script and functions with examples.

Table 5.7 Symbols and markers for plotting

Symbols	Lines	Symbols	Colors	Symbols	Markers
_	Solid	R	Red	*	Asterisk
_ _	Dashed	B	Blue	+	Plus
. . .	Doted	G	Green	O	Circle
-.-.-	Dash-doted	C	Cyan	X	Cross
		M	Magenta	.	Point
		Y	Yellow	D	Diamond
		B	Black	S	Square

(a) MatLab Script

When there are a number of commands and all are being executed for multiple times with different or same data, then it is better to create a MatLab script. MatLab script is the set of commands and variables written inside a single file, with extension of ".m". To create and edit these file, MatLab editor is used. These can also be easily executed by pressing the run button in editor.

Example of script:

```
script1.m
% commented statement
% simple addition
clc;
clear all;

a=10;
b=20;
c=a+b;
disp(c);
```

(b) MatLab User-Defined Functions

Similar to MatLab scripts, MatLab function files are also saved with extension ".m". But there is a proper syntax for defining the functions inside the function file. To create the function file, again MatLab editor is used. The first line of the function file starts with the keyword "function," then name of return variable, name of function, and its input arguments or parameters. The name of the function file must be the same as the name of the function, otherwise it will not work.

Example of a function file:

```
addition.m
function y = addition(x1,x2)
y=x1+x2;
```

To use this function, similar way is adopted as in the case of MatLab in-built functions. Simply type the name of the function with input arguments, they may be variable or direct values.

Direct execution:

```
>>addition(10,20)
```

ans =

   30

Execution with variables

```
>>x1=10
```

x1 =

   10

```
>>x2=20
```

x2 =

   20

```
>> y=addition(x1, x2);
>>y
```

y =

   30

Example of a function with two or more output variables:

calculation.m

```function [y1 y2]=calculation(x1,x2)```
 ```y1=x1+x2;```
 ```y2=x1-x2;```

Execution in similar manner,

```
>> [y1 y2]=calculation(x1, x2);
>> y1
```

y1 =

   30

```
>> y2
```

y2 =

   -10

Exercise for functions: try to build the function file for area of circle, calculator with +, -, *, and / operations.

5.2 Conditional Statements and Loops

Similar to other programming languages, MatLab also provides the basic conditional statement, loops, and logical operators. In this section, all the basic conditional statements and loops are demonstrated with examples. In the examples given below, just by putting different values of variables *a* and *b*, you can observe their working.

(a) IF, IF ELSE, IF ELSEIF

```
example_cond_1.m
```
```
clc;
clear all;
a=-1;
if (a == 0)
    fprintf('\nInside IF')
    fprintf('\na=%d',a)
elseif (a < 0)
    fprintf('\nInside ELSEIF')
    fprintf('\na=%d',a)
else
    fprintf('\nInside ELSE')
    fprintf('\na=%d',a)
end
```

Run the above script as it is, and then put a = −1 for the next time. You can also try other values.

```
example_cond_2.m
```
```
clc;
clear all;
a=1;
b=-1;
if (a == 0 && b ==0)
    fprintf('\nInside IF with AND')
    fprintf('\na=%d b=%b',a,b)
elseif (a == 0 || b > 0)
    fprintf('\nInside ELSEIF_1 with OR')
    fprintf('\na=%d b=%d',a,b)
elseif (b ~= 0)
    fprintf('\nInside ELSEIF_2 with NOT')
    fprintf('\na=%d b=%d',a,b)
else
    fprintf('\nInside ELSE')
    fprintf('\na=%db=%d',a,b)
end
```

```
example_while_1.m
clc;
clear all;
n=0;
while (n<10)
     n=n+1;
     fprintf('%d \t', n )
end
```

Try values a = 1 and b = −1 for the execution of NOT portion. Try execution of the above examples with different values of a and b.

(b) WHILE and FOR Loop

```
example_for_1.m
clc;
clear all;
n=0;
fprintf('i    n')
for i=1:5
     n=n+i;
     fprintf('\n%d %d', i, n)
end
```

Output of the above code is:

```
>> 1   2   3   4   5   6   7   8   9   10
```

```
example_for_2.m
clc;
clear all;
n=0;
fprintf('i    n')
for i=1:2:10
     n=n+i;
     fprintf('\n%d %d', i, n)
end
```

Output of the above code is:

```
>>i  n
1 1
2 3
3 6
4 10
5 15
```

Output of the above code is:

```
>>i  n
1 1
3 4
5 9
7 16
9 25
```

---

example_for_3.m

```
clc;
clear all;
n=0;
A=[23 34 44 45 56 78 99 12];
fprintf('i    A    n')
for i=1:8
    n=n+A(i);
    fprintf('\n%d   %d   %d', i, A(i), n)
end
```

Output of the above code is:

```
>>  i A n
1 23 23
2 34 57
3 44 101
4 45 146
5 56 202
6 78 280
7 99 379
8 12 391
```

---

example_for_4.m

```
clc;
clear all;
B=[1 2 3; 4 5 6; 7 8 9];
fprintf('i   j   B')
for i=1:3
    for j=1:3
        fprintf('\n%d   %d   %d', i, j, B(i,j))
    end
end
```

Output of the above code is:

```
>>i  j  B
1  1  1
1  2  2
1  3  3
2  1  4
2  2  5
2  3  6
3  1  7
3  2  8
3  3  9
```

Example to break FOR loop with the help of WHILE, in case if element inside the array A is negative.

```
example_for_while_1.m
clc;
clear all;
A=[1 2 3 0 0 5  1 -6 7 9 1];
for i=1:length(A)
while (A(i)<0)
    fprintf('Index: %d Value: %d',i, A(i))
    break
end
end
```

Output of the above code is:

```
>>Index: 8 Value: -6
```

5.3 Fundamental Concepts of Programming for Signal/ Image Processing

In this section, fundamental steps to begin the implementation of signal processing methods have been demonstrated. Here, signal loading, plotting, and preprocessing are demonstrated. In this example, sample electroencephalogram (EEG) signals are loaded from .mat file. Name of the file is EEG1.m. These signals are taken from DEAP dataset of EEG signals [4, 5].

5.3.1 Load and Plotting of 1D Data (Signals)

load_plot_data.m

```
clc; % To clear screen
clear all; % To clear workspace
close all; % To close all figures and windows

load EEG1 % To load EEG1.mat file having EEG
signals

[ch samples]=size(EEG1); % To check size of EEG
data
fprintf('\n channels :%d',ch); % Row count as
number of channels
fprintf('\n samples :%d',samples); % Column count
as number of samples


D =  5; % signal duration (sec) for plotting
fs = 128; % sampling rate (samples per sec)
T = 1/fs; % sampling period, time interval for
recording of samples
t = [T:T:D]; % time vector used for plotting of X
or time axis
Total_Duration = samples/fs;

fprintf('\n recording Time in Sec :%f',
Total_Duration); % Total duration of signal

figure
plot(t,EEG1(1,1:640)); % Ploting 5 sec data
(640=5*128)
xlabel('Time (Sec)')
ylabel('Amplitude (\mu V)')
```

Output of the above code is given below with the plot of data in Fig. 5.6.

```
channels :32
samples :7680
recording Time in Sec :60.000000
```

For plotting of different channels at the same time, figure command can be used. With the help of figure command, in the same script, multiple figures can be plotted. To plot multiple signals (channels) in a single figure, subplot

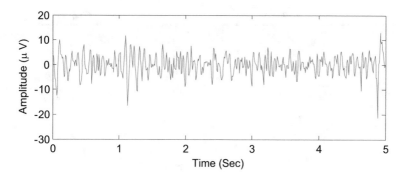

Fig. 5.6 Plot of EEG signal

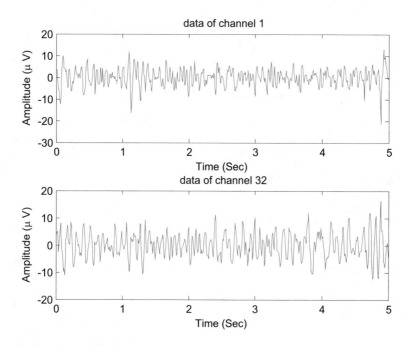

Fig. 5.7 Plot of two EEG signals

command can used, which is demonstrated below, these line of codes simply added to the above script. Plot of multiple signals on the same figure is shown in Figs. 5.7 and 5.8.

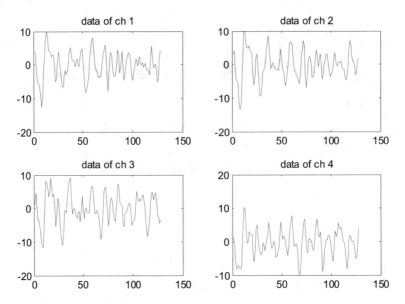

Fig. 5.8 Sub-plots of multiple EEG signals

line of codes 1

```
figure
subplot(2,1,1)
plot(t,EEG1(1,1:640));
title('data of channel 1')
xlabel('Time (Sec)')
ylabel('Amplitude (\mu V)')
subplot(2,1,2)
plot(t,EEG1(32,1:640));
title('data of channel 32')
xlabel('Time (Sec)')
ylabel('Amplitude (\mu V)')
```

lines of codes 2

```
figure
subplot(2,2,1)
plot(EEG1(1,1:128))
title('data of ch 1')
subplot(2,2,2)
plot(EEG1(2,1:128))
title('data of ch 2')
subplot(2,2,3)
plot(EEG1(3,1:128))
title('data of ch 3')
subplot(2,2,4)
plot(EEG1(4,1:128))
title('data of ch 4')
```

5.3.2   Filtering over Data

In code given below, filtering of EEG signals is demonstrated. Several finite impulse response (FIR) filters are created, and signal is filtered in different frequency bands. Later on, filtered signal in specific frequency band can be used for study. Figures 5.9 and 5.10 show the output filtered signals and their frequency spectrums, respectively.

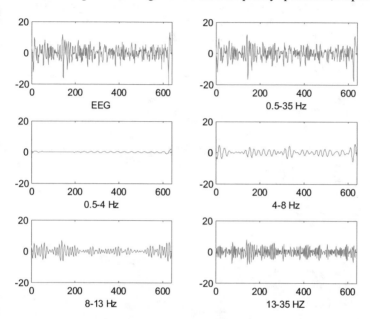

Fig. 5.9 Plot of filtered signals in different frequency bands

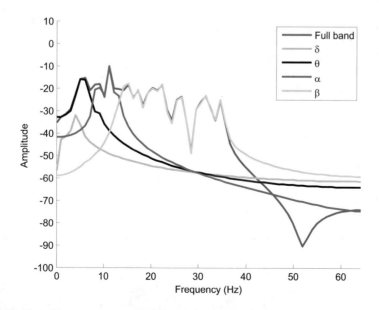

Fig. 5.10 Plot of frequency spectrum of filtered signals of specific frequency bands

```
filtering1d.m
```

```
clc;
clear all;
close all;

D =  5; %signal duration
fs = 128; % sampling rate, i.e. N points pt
sec used to represent sine wave
T = 1/fs; % sampling period, i.e. for this
e.g. points at 1 ms intervals
t = [T:T:D]; % time vector

load EEG1;
data=EEG1(1,1:640); % taking 5 sec data
(640=5*128)
% conventioanl frequency bands of EEG
fp0=[0.5 35];
fp1=[0.5 4];
fp2=[4 8];
fp3=[8 13];
fp4=[13 35];

% conversion into angular frequency
wp0=(2/fs).*fp0;
wp1=(2/fs).*fp1;
wp2=(2/fs).*fp2;
wp3=(2/fs).*fp3;
wp4=(2/fs).*fp4;

% Calculation of filter coefficients with FIR
1 filter
% Number of coeffecients (200)
b0=fir1(200,wp0);
b1=fir1(200,wp1);
b2=fir1(200,wp2);
b3=fir1(200,wp3);
b4=fir1(200,wp4);

%Plotting the filter response

figure
freqz(b0,1,500,fs);
hold
```

```
freqz(b1,1,500,fs);
freqz(b2,1,500,fs);
freqz(b3,1,500,fs);
freqz(b4,1,500,fs);
TITLE('Magnitude and Phase response');

y0=conv(data,b0);
y0=y0((((size(b0,2))/2)+1):end-
((size(b0,2))/2));

y1=conv(data,b1);
y1=y1((((size(b1,2))/2)+1):end-
((size(b1,2))/2));

y2=conv(data,b2);
y2=y2((((size(b2,2))/2)+1):end-
((size(b2,2))/2));

y3=conv(data,b3);
y3=y3((((size(b3,2))/2)+1):end-
((size(b3,2))/2));

y4=conv(data,b4);
y4=y4((((size(b4,2))/2)+1):end-
((size(b4,2))/2));

figure
subplot(3,2,1); plot(data); xlabel('EEG');
axis([0 640 -20 20])
subplot(3,2,2); plot(y0); xlabel('0.5-35 Hz');
axis([0 640 -20 20])
subplot(3,2,3); plot(y1); xlabel('0.5-4 Hz');
axis([0 640 -20 20])
subplot(3,2,4); plot(y2); xlabel('4-8 Hz');
axis([0 640 -20 20])
subplot(3,2,5); plot(y3); xlabel('8-13 Hz');
axis([0 640 -20 20])
subplot(3,2,6); plot(y4);xlabel('13-35 HZ');
axis([0 640 -20 20])

figure
L = length(data);
myFFT = fft(data,fs);
myFFT0 = fft(y0,fs);
```

```
myFFT1 = fft(y1,fs);
myFFT2 = fft(y2,fs);
myFFT3 = fft(y3,fs);
myFFT4 = fft(y4,fs);


myFFT=myFFT/L; % Normalizing
myFFT0=myFFT0/L;
myFFT1=myFFT1/L;
myFFT2=myFFT2/L;
myFFT3=myFFT3/L;
myFFT4=myFFT4/L;
freq = fs/2*linspace(0,1,fs/2);%create a range
with all frequency values


% Plot single-sided amplitude spectrum.
hold
plot(freq,db(abs(myFFT0(1:length(freq)))),'r',
'LineWidth',2);
axis([0 64 -100 10])
plot(freq,db(abs(myFFT1(1:length(freq)))),'g',
'LineWidth',2);
plot(freq,db(abs(myFFT2(1:length(freq)))),'k',
'LineWidth',2);
plot(freq,db(abs(myFFT3(1:length(freq)))),'m',
'LineWidth',2);
plot(freq,db(abs(myFFT4(1:length(freq)))),'c',
'LineWidth',2);
legend('Full
band','\delta','\theta','\alpha','\beta')
xlabel('Frequency (Hz)');
ylabel('Amplitude ');
```

5.3.3   Loading and Plotting 2D Data (Images)

Image as 2D or 3D data matrix is represented in several ways. Binary images having values 0 and 1 as $m \times n$ matrix, where m is the number of rows and n is the number of columns (both represent number of pixels or image resolution). Gray-level images have pixel values 0–255 in $m \times n$ matrix. In RGB images, pixel values are from 0 to 255 and stored as $m \times n \times 3$. This $m \times n \times 3$ is a 3D matrix of three planes of red, green, and blue, respectively, each of $m \times n$ size. Some multidimensional images having values 0–255 are stored as $m \times n \times p$ matrix of p number of plane or layers. Below is the simple exercise of loading, plotting (display) and analysis of images.

5.3.4   Load and Display Image

In the following code, example of loading image and how to gain basic information of image is demonstrated. Figure 5.11 shows the image as output with some basic information of size and dimensions. This image is taken from [6, 7].

```
example_image_1.m

clc;
clear all;
close all;

data=imread('PET.jpg');
[m n l]=size(data);
fprintf('rows = %d  columns = %d  layers = %d',
m, n, l);
imshow(data);
xlabel('PET Image of Brain')
```

Output of above code is:

```
>>rows = 339 columns = 300 layers = 3
```

Fig. 5.11 PET image of brain loaded from .jpg format

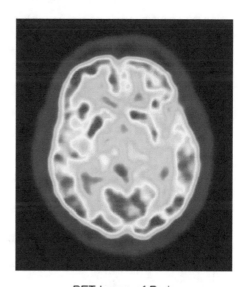

PET Image of Brain

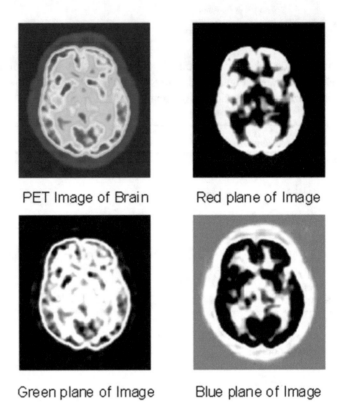

Fig. 5.12 Different planes of PET image

As the above image is a color image having red, green, and blue layers or planes. In the following code, each plane is shown separately. Figure 5.12 shows different planes of the same image using subplots. Minimum and maximum pixel values of each plane are also printed in the output.

```
example_image_2.m
clc;
clear all;
close all;
data=imread('PET.jpg');
[m n l]=size(data);
fprintf('rows = %d   columns = %d   layers = %d',
m, n, l);
subplot(2,2,1), imshow(data); xlabel('PET Image
of Brain');
subplot(2,2,2), imshow(data(:,:,1)); xlabel('Red
plane of Image');
```

```
subplot(2,2,3), imshow(data(:,:,2)); xla-
bel('Green plane of Image');
subplot(2,2,4), imshow(data(:,:,3)); xla-
bel('Blue plane of Image');

% min and max pixel value of each plane
Rmin = min(min(data(:,:,1)));
Gmin = min(min(data(:,:,2)));
Bmin = min(min(data(:,:,3)));
Rmax = max(max(data(:,:,1)));
Gmax = max(max(data(:,:,2)));
Bmax = max(max(data(:,:,3)));

fprintf('\n Rmin    Rmax    Gmin    Gmax    Bmin
Bmax');
fprintf('\n %d       %d      %d      %d      %d      %d
',Rmin,Rmax,Gmin,Gmax,Bmin,Bmax);
```

```
>>rows = 339 columns = 300 layers = 3
Rmin Rmax Gmin Gmax Bmin Bmax
0     255  0    255  0    255
```

Conversion of image format and its color model can also be performed very easily in MatLab. The following code converts the color image in gray-level image. Figure 5.13 shows the converted gray-level image. In output, minimum and maximum gray-level values are also printed.

Fig. 5.13 Converted gray-level image of color PET image

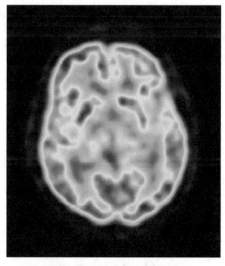

Gray PET Image of Brain

example_image_3.m

```
clc;
clear all;
close all;

data=imread('PET.jpg');
data1=rgb2gray(data); % RGB to Gray scale Image
[m n l]=size(data1);
fprintf('rows = %d  columns = %d  layers = %d',
m, n, l);

imshow(data1); xlabel('Gray PET Image of Brain');

% min and max pixel value of each plane
gray_min = min(min(data1(:,:,1)));
gray_max = max(max(data1(:,:,1)));

fprintf('\n min  max');
fprintf('\n %d     %d',gray_min, gray_max);
```

The output of the above code is:

```
>>rows = 339 columns = 300 layers = 1
  min max
  14   228
```

The following code, demonstrated the difference between gray-level and black and white images. Here, imtool command is used to show the images. The advantage of this command is that, on placing the mouse pointer over image, it shows the pixel value at left bottom of the window in which image is shown. This will be very useful for observing and analyzing the image and its different regions. Figures 5.14, 5.15, and 5.16 show the imtool window with different color model images. Position of mouse pointer is shown by white and red color arrow.

example_image_4.m

```
clc;
clear all;
close all;
data=imread('PET.jpg');
data1=rgb2gray(data); % RGB to Gray scale Image
data2=im2bw(data); % RGB to Binay Image (Black &
White)
imtool(data);
imtool(data1);
imtool(data2);
```

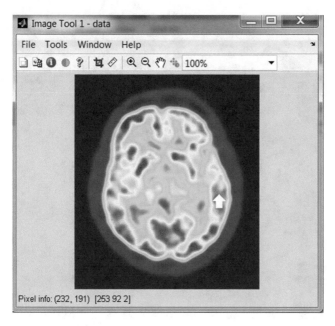

Fig. 5.14 Imtool window with color PET image

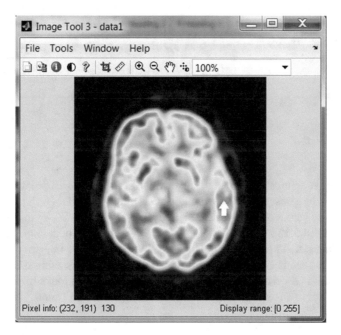

Fig. 5.15 Imtool window with gray PET image

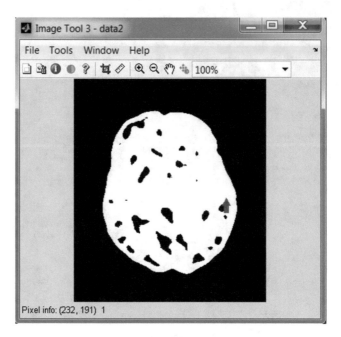

Fig. 5.16 Imtool window with black and white PET image

Fig. 5.17 Data matrix of
different color model
images

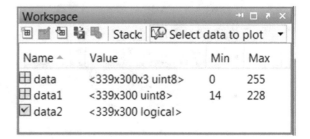

In the above figures, each time the pixel value is shown in a different way. Proper
names for each representation shown in Fig. 5.14 are: Pixel info: (X, Y) [R G B] that
is the intensity values of red, green, and blue plane at positing (X, Y) in case of color
images. In Figs. 5.15 and 5.16, Pixel info: (X, Y) [intensity] in case of gray images
and Pixel info: (X, Y) [binary value] in case of black and white images, respectively.
This concept becomes clearer by observing the work space having data matrix of
these images, as shown in Fig. 5.17. For color image, matrix looks like "data". For
gray and binary (black and white) images, matrix look like "data1" and "data2",
respectively.

5.3.5   Filtering of Images

In this section, the concept of image filtering is demonstrated. A simply way to apply filters over images and check their effects in spatial and frequency domains are also demonstrated. In below listed code, simple low pass and high pass filters are used. These are simple concepts of averaging and edge detection in images. Figures 5.18 and 5.19 show the effect of filters on gray image in spatial and frequency domains.

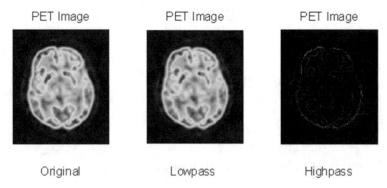

Fig. 5.18   Original, low-pass filtered and high-pass filtered gray PET images in spatial domain

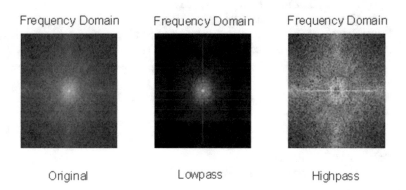

Fig. 5.19   Original, low-pass filtered and high-pass filtered gray PET images in frequency domain

example_image_5.m

```
clc;
clear all;
close all;

data=imread('PET.jpg');
data1=rgb2gray(data); % RGB to Gray scale Image
[m n l] = size(data1);

% Display the original gray scale image.
subplot(2, 1, 1);
imshow(data1);

% low pass filter (smoothing by mean filter)
lp=[1/9 1/9 1/9; 1/9 1/9 1/9; 1/9 1/9 1/9];
data1lp = filter2(lp, im2double(data1));

% high pass filter (edges only)
hp=[-1 -1 -1; -1 8 -1; -1 -1 -1];
data1hp = filter2(hp, im2double(data1));

subplot(1,3,1), imshow(data1); xla-
bel('Original'); title('PET Image')
subplot(1,3,2), imshow(data1lp); xla-
bel('Lowpass'); title('PET Image')
subplot(1,3,3), imshow(data1hp); xla-
bel('Highpass'); title('PET Image')

% In frequency domain
Fd=fft2(data1);
Sd=fftshift(log(1+abs(Fd)));
Fdlp=fft2(data1lp);
Sdlp=fftshift(log(1+abs(Fdlp)));
Fdhp=fft2(data1hp);
Sdhp=fftshift(log(1+abs(Fdhp)));

figure
subplot(1,3,1), imshow(Sd, []); xla-
bel('Original'); title ('Frequency Domain')
subplot(1,3,2), imshow(Sdlp,[]); xla-
bel('Lowpass'); title ('Frequency Domain')
subplot(1,3,3), imshow(Sdhp,[]); xla-
bel('Highpass'); title ('Frequency Domain')
```

Fig. 5.20 Original, low-pass filtered and high-pass filtered gray sample images in spatial domain

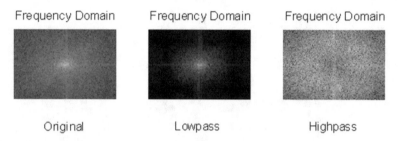

Fig. 5.21 Original, low-pass filtered and high-pass filtered gray sample images in frequency domain

The same code is executed again, but with different sample image taken from [8, 9]. Figures 5.20 and 5.21 show the output with low-pass and high-pass filtered sample images in spatial and frequency domains. By doing this, one can get more clear understanding. Filter parameters and its size (values of *hp* and *lp* matrix in code) can also be changed.

References

1. https://in.mathworks.com/help/matlab/
2. https://www.mathworks.com/support.html
3. D. Houcque, Introduction to MATLAB For Engineering Students, version 1.2 (Northwestern University, August 2005)
4. https://www.eecs.qmul.ac.uk/mmv/datasets/deap/
5. S. Koelstra et al., DEAP: A database for emotion analysis; using physiological signals. IEEE Transactions on Affective Computing **3**(1), 18–31 (2012). https://doi.org/10.1109/T-AFFC. 2011.15
6. http://large.stanford.edu/courses/2015/ph241/krishnamurthi1/
7. https://commons.wikimedia.org/wiki/File:PET-image.jpg
8. J. Li, J.Z. Wang, Automatic linguistic indexing of pictures by a statistical modeling approach. IEEE Transactions on Pattern Analysis and Machine Intelligence **25**(9), 1075–1088 (2003)
9. http://wang.ist.psu.edu/docs/related/

Index

Printed in the United States
by Baker & Taylor Publisher Services